Important Information

This book belongs to: _____

_____

_____

_____

phone: _____

PE examination date: _____ time: _____

location: _____

_____

_____

Examination Board Address: _____

_____

_____

phone: _____

Application was requested on this date: _____

Application was received on this date: _____

Application was accepted on this date: _____

Information received during the examination (such as booklet numbers) follows:

_____

_____

"Why do engineers prefer GLP study guides?"

"The multiple-choice and essay sample problems in your PE review were exactly like what I found when I took the actual exam!"
—*John Jurgensen, Louisiana*

•

"In your review books you cut to the chase. Yours is the only review that doesn't scare the pants off people. It's exactly what I want for teaching my review course."
—*John Thorington*

•

"I passed the first time I took the test. Potter's book is a great, great review. It's right to the point and doesn't include extra stuff you don't need. All I did was work through the book. My friends took review courses and didn't use your book—they failed! Do you have a PE review?"
—*Dana, calling to order our PE Civil review*

•

"I used your FE book and thought it was far superior to any other. I took the exam and did great. Now I need your PE Review!"
—*Tom Roach, Indiana Midwest Steel*

•

"Your *Principles & Practice of Civil Engineering* review was well worth the money spent! It was extremely easy to use during the exam and greatly increased my comfort level. The chapters on soils, hydrology, and open channel flow are exceptional."   —*Mark MacIntire, Virginia*

•

"When you're working 50 hours a week, you don't have a lot of extra time to prepare. Potter books are condensed and focus on only what is on the exam. It's a good strategy to concentrate on what is on the exam—just what you need to pass. It doesn't overwhelm you."
—*James Alt, graduate of University of Wisconsin*

•

"I used your FE review the second time I took the exam and finally passed it! So, obviously I turned to your PE review and was equally pleased. It is informative, to-the-point, and covers all the material needed for the exam. Unlike other reviews, yours are easy to follow and I don't have to dig through anything that is not on the exam."
—*Gena Swift, Illinois*

International Standard Book Number 1-881018-21-0

Second Edition, Copyright © 1998 by Great Lakes Press, Inc.

All comments and inquiries should be addressed to:
    Great Lakes Press—Customer Service
    PO Box 550
    Wildwood, MO 63040-0550
    Phone: (314) 273-6016
    Fax: (314) 273-6086
    Website: www.glpbooks.com
    Email: custserv@glpbooks.com

Library of Congress
Cataloging-in-Publication Data

Printed in the USA by Braun-Brumfield, Inc., Ann Arbor, Michigan.

10 9 8 7 6 5 4 3 2 1

...from the Professors who know it best...

PRINCIPLES & PRACTICE OF CIVIL ENGINEERING

Solutions Manual

...to the Practice Problems of each chapter...

Editor: MERLE C. POTTER, PhD, PE
Professor, Michigan State University

Authors:		
	Mackenzie L. Davis, PhD, PE	Water Quality
	Richard W. Furlong, PhD, PE	Structures
	David A. Hamilton, MS, PE	Hydrology
	Ronald Harichandran, PhD, PE	Structures
	Frank Hatfield, PhD, PE	Economics
	Thomas L. Maleck, PhD, PE	Transportation
	George E. Mase, PhD	Mechanics
	Merle C. Potter, PhD, PE	Fluid Mechanics
	David C. Wiggert, PhD, PE	Hydraulics
	Thomas F. Wolff, PhD, PE	Soils

The authors are professors at Michigan State University, with
the exception of R. W. Furlong, who teaches at the University of
Texas at Austin and D. A. Hamilton who is employed by the
Michigan Department of Natural Resources.

published by:

GREAT LAKES PRESS
P.O. Box 550
Wildwood, MO 63040-0550
Customer Service (314) 273-6016
Fax (314) 273-6086
www.glpbooks.com   custserv@glpbooks.com

Table of Contents

Solutions to Practice Problems

Preface

This solutions manual provides the solutions to the practice problems at the end of each chapter in the *Principles & Practice of Civil Engineering Review*, second edition. Most of the solutions are intentionally not as detailed as the examples in the book; they are intended to provide only the primary steps in each solution showing the appropriate numbers substituted into the required equation resulting in the answer. Units are sometimes displayed if they are not obvious. Occasionally, subtle points may be discussed. In general, some knowledge of a problem may be needed to fully understand all the steps presented; this knowledge is assumed to come from the presentation and the examples worked out in detail in the review material.

The degree of difficulty and length of solution for each problem varies considerably. Some are relatively easy and others are quite difficult. This parallels the actual exam. Typically, the easier problems come first, especially in the multiple-choice questions.

The examples in the book and the solutions in this manual have been carefully solved with the hope that errors have not been introduced. Even though extreme care is taken and problems are worked repeatedly, errors do creep in. We would appreciate knowing about any errors that you may find. They can be eliminated in future printings. Suggestions for changes and improvements are also welcomed. Write to: GLP, POB 550, Grover, MO 63040, send email to: custserv@glpbooks.com, fax: 314-273-6086.

Why a Separate Solutions Manual?

It is necessary to place the solutions in a separate book because some states do not allow solutions to exam-like problems into the exam room, where our review book is meant to be used. One can either check with one's state board to decide if this manual is allowed in the exam room, or it can be left outside the exam room if it is not allowed. Such decisions depend on individual interpretations of proctors and often vary.

Study Strategy

If one is not familiar with the subjects of a sufficient number of chapters, it may be advisable to enroll in a review course where the material is presented by experts. Such a review course may be needed for a "pass" on the exam.

Since one selects four problems to work from a set of 12 problems in both the a.m. and p.m., it is not necessary to become proficient in all subjects included in the exam. We recommend that at least six major areas are studied, assuming that the required problems can be selected from these areas.

Dr. Merle C. Potter
Okemos, Michigan

Mathematics

by Merle C. Potter

Chapter 1

Solutions to Practice Problems

1.1 c) $100 = 10e^{2t}$. $\therefore \ln e^{2t} = \ln 10$. $\therefore 2t = 2.303$ $\therefore t = 1.151$

1.2 b) $\ln x = 3.2$. $\therefore e^{3.2} = x$. $\therefore x = 24.53$

1.3 c) $\ln_5 x = -1.8$. $\therefore 5^{-1.8} = x$. $\therefore x = 0.0552$

1.4 a) $x = \dfrac{-(-2) \pm \sqrt{2^2 - 4(3)(-2)}}{3 \cdot 2} = 1.215$

1.5 e) $(4+x)^{\frac{1}{2}} = 4^{1/2} + \dfrac{1}{2} 4^{-1/2} x + \dfrac{\frac{1}{2}\left(1 - \frac{1}{2}\right)}{2} 4^{-3/2} x^2 + \cdots = 2 + x/4 - x^2/64 + \cdots$

1.6 a) $\dfrac{2}{x\left(x^2 - 3x + 2\right)} = \dfrac{A_1}{x} + \dfrac{A_2}{x-2} + \dfrac{A_3}{x-1} = \dfrac{A_1\left(x^2 - 3x + 2\right) + A_2\left(x^2 - x\right) + A_3\left(x^2 - 2x\right)}{x\left(x^2 - 3x + 2\right)}$

$\therefore \left.\begin{array}{r} A_1 + A_2 + A_3 = 0 \\ -3A_1 - A_2 - 2A_3 = 0 \\ 2A_1 = 2 \end{array}\right\} \quad \left.\begin{array}{r} A_1 = 1 \\ A_2 + A_3 = -1 \\ -A_2 - 2A_3 = 3 \end{array}\right\} \quad \begin{array}{l} A_3 = -2 \\ A_2 = 1 \end{array}$

1.7 b) $\dfrac{4}{x^2\left(x^2-4x+4\right)}=\dfrac{A_1}{x}+\dfrac{A_2}{x^2}+\dfrac{A_3}{x-2}+\dfrac{A_4}{(x-2)^2}$

$$=\dfrac{A_1\left(x^3-4x^2+4x\right)+A_2\left(x^2-4x+4\right)+A_3\left(x^3-2x^2\right)+A_4x^2}{x^2(x-2)^2}$$

$\therefore$
$\left.\begin{array}{r}A_1+A_3=0\\-4A_1+A_2-2A_3+A_4=0\\4A_1-4A_2=0\\4A_2=4\end{array}\right\}$
$\begin{array}{l}A_2=1\\A_1=1\\A_3=-1\\A_4=1\end{array}$

1.8 d) at $t=0$, population $=A$. $\therefore 2A=Ae^{0.4t}$. $\ln 2=0.4t$. $\therefore t=1.733$

1.9 a) $\sin\theta=0.7$ $\therefore \theta=44.43°$. $\tan 44.43°=0.980$

1.10 e) $\tan\theta=5/7$. $\therefore \theta=35.54°$

1.11 c) $\tan\theta\sec\theta\left(1-\sin^2\theta\right)\big/\cos\theta=\tan\theta\dfrac{1}{\cos\theta}\cos^2\theta\dfrac{1}{\cos\theta}=\tan\theta$

1.12 d) $3^2=4^2+2^2-2\cdot2\cdot4\cos\theta$. $\therefore \cos\theta=0.6875$. $\theta=46.6°$. $\therefore$ rad $=0.813$

1.13 a) $L^2=850^2+732^2-2\cdot850\cdot732\cos154°$. $\therefore L=1542$ m

1.14 e) $\cos2\theta=\cos^2\theta-\sin^2\theta=1-\sin^2\theta-\sin^2\theta=1-2\sin^2\theta$. $\therefore 2\sin^2\theta=1-\cos2\theta$

1.15 b) $\theta=\pi(n-2)/n=\pi(8-2)/8$ radians. $\dfrac{6\pi}{8}\times\dfrac{180}{\pi}=135°$

1.16 a) Area $=$ Area$_{top}+$ Area$_{sides}=\pi R^2+\pi DL$

$$=\pi\times7.5^2+\pi\times15\times10=648\text{ m}^2.\quad 648\div10\approx65$$

1.17 a) $y=mx+b$. $y=-2x+b$. $0=-2(2)+b$. $\therefore b=4$. $\therefore y=-2x+4$

1.18 b) $y=mx+b$. $0=4m+b$. $-6=b$. $\therefore m=3/2$. $\therefore y=3x/2-6$ or $2y=3x-12$

1.19 d) $3x-4y-3=0$. $A=3, B=-4$. $d=\dfrac{|3\times6-4\times8-3|}{\sqrt{3^2+(-4)^2}}=3.4$

1.20 c) $B^2-AC=2^2-1\times4=0$. $\therefore$ parabola

1.21 b) $xy=\pm k^2=4$

1.22 c) $\dfrac{x^2}{25^2}+\dfrac{y^2}{50^2}=1$. $\therefore 4x^2+y^2=2500$

1.23 **b)** $x = r\cos\theta = 5\times 0.866 = 4.33.$ $y = r\sin\theta = 5\times 0.5 = 2.5$

Spherical coordinates:

$$r = \sqrt{4.33^2 + 2.5^2 + 12^2} = 13. \quad \phi = \cos^{-1}\frac{z}{r} = \cos^{-1}\frac{12}{13} = 22.6°.$$

$$\theta = \tan^{-1}\frac{y}{x} = \tan^{-1}\frac{2.5}{4.33}. \quad \therefore \theta = 30°.$$

1.24 **c)** **a)** is rectangular coordinates. **b)** is spherical coordinates.

1.25 **a)** $\dfrac{3-i}{1+i} = \dfrac{3-i}{1+i}\dfrac{1-i}{1-i} = \dfrac{3-1-4i}{1-(-1)} = \dfrac{1}{2}(2-4i) = 1-2i$

1.26 **d)** $1+i = re^{i\theta}.$ $r = \sqrt{1^2 + 1^2} = \sqrt{2}.$ $\theta = \tan^{-1}\dfrac{1}{1} = \pi/4$ rad. $\therefore 1+i = \sqrt{2}e^{i\pi/4}.$

$$\therefore (1+i)^6 = \left(\sqrt{2}\right)^6 e^{i\pi 3/2} = 8\left(\cos\frac{3\pi}{2} + i\sin\frac{3\pi}{2}\right) = -8i.\ \text{Note: angles are in radians.}$$

1.27 **b)** $1+i = \sqrt{2}e^{i\pi/4}.$ $\therefore (1+i)^{1/5} = 1.414^{1/5}e^{i\pi/20} = 1.072\left(\cos\dfrac{\pi}{20} + i\sin\dfrac{\pi}{20}\right)$

$$= 1.06 + 0.168i$$

1.28 **c)** $(3+2i)(\cos 2t + i\sin 2t) + (3-2i)(\cos 2t - i\sin 2t) = 6\cos 2t + 4i(i\sin 2t)$

$$= 6\cos 2t - 4\sin 2t$$

1.29 **e)** $6(\cos 2.3 + i\sin 2.3) - 5(\cos 0.2 + i\sin 0.2) = 6(-0.666 + 0.746i) - 5(0.98 + 0.199i)$

$$= -8.90 + 3.48i$$

$$= 9.56 \angle 158.6°$$

1.30 **e)** $\begin{vmatrix} 3 & 2 & 1 \\ 0 & -1 & -1 \\ 2 & 0 & 2 \end{vmatrix} = -6 - 4 + 2 = -8$

1.31 **a)** $\begin{vmatrix} 1 & 0 & 1 & 1 \\ 2 & -1 & 0 & 1 \\ 0 & 0 & 2 & 0 \\ 3 & 2 & 1 & 1 \end{vmatrix} = 2\begin{vmatrix} 1 & 0 & 1 \\ 2 & -1 & 1 \\ 3 & 2 & 1 \end{vmatrix} = 2(-1 + 4 + 3 - 2) = 8$

Note : Expand using the third row.

1.32 **b)** $(-1)^3 \begin{vmatrix} 2 & 1 \\ 0 & 2 \end{vmatrix} = -4$

1.33 **e)** $(-1)^7 \begin{vmatrix} 1 & 0 & 1 \\ 2 & -1 & 0 \\ 3 & 2 & 1 \end{vmatrix} = -(-1 + 4 + 3) = -6$

1.34 **d)** $\left[a_{ij}\right]^{+} = \begin{bmatrix} A_{11} & A_{21} \\ A_{12} & A_{22} \end{bmatrix} = \begin{bmatrix} 2 & 4 \\ 0 & 1 \end{bmatrix}$

1.35 **a)** $\left[a_{ij}\right]^{-1} = \dfrac{\left[a_{ij}\right]^{+}}{\left|a_{ij}\right|} = \dfrac{\begin{bmatrix} 1 & -3 \\ -1 & 2 \end{bmatrix}}{-1} = \begin{bmatrix} -1 & 3 \\ 1 & -2 \end{bmatrix}$

1.36 **b)** $\begin{bmatrix} 2 & -1 \\ 3 & 2 \end{bmatrix}\begin{bmatrix} 2 \\ 1 \end{bmatrix} = \begin{bmatrix} 4-1 \\ 6+2 \end{bmatrix} = \begin{bmatrix} 3 \\ 8 \end{bmatrix}$

1.37 **a)** $\begin{bmatrix} 1 & 2 \\ 2 & 1 \end{bmatrix}\begin{bmatrix} -1 & 0 \\ 1 & 2 \end{bmatrix} = \begin{bmatrix} 1 & 4 \\ -1 & 2 \end{bmatrix}$

1.38 **d)** $\left[a_{ij}\right] = \begin{bmatrix} 3 & 2 & 0 \\ 1 & -1 & 1 \\ 4 & 0 & 2 \end{bmatrix}.$ $\left[a_{ij}\right]^{+} = \begin{bmatrix} -2 & -4 & 2 \\ 2 & 6 & -3 \\ 4 & 8 & -5 \end{bmatrix}.$ $\left|a_{ij}\right| = -2.$

$\therefore \left[a_{ij}\right]^{-1} = \dfrac{\left[a_{ij}\right]^{+}}{\left|a_{ij}\right|} = \begin{bmatrix} 1 & 2 & -1 \\ -1 & -3 & 3/2 \\ -2 & -4 & 5/2 \end{bmatrix}.$ $\left[x_j\right] = \left[a_{ij}\right]^{-1}\left[r_i\right] = \left[a_{ij}\right]^{-1}\begin{bmatrix} -2 \\ 0 \\ 4 \end{bmatrix} = \begin{bmatrix} -6 \\ 8 \\ 14 \end{bmatrix}$

1.39 **a)** $\begin{vmatrix} 1-\lambda & 2 \\ 3 & 2-\lambda \end{vmatrix} = \lambda^2 - 3\lambda + 2 - 6 = \lambda^2 - 3\lambda - 4 = 0.$

$(\lambda - 4)(\lambda + 1) = 0.$ $\therefore \lambda = 4, -1$

1.40 **b)** Use $\lambda = 4.$ $\begin{bmatrix} -3 & 2 \\ 3 & -2 \end{bmatrix}\begin{bmatrix} x_1 \\ x_2 \end{bmatrix} = 0.$

$\begin{array}{l} -3x_1 + 2x_2 = 0. \\ 3x_1 - 2x_2 = 0. \end{array}$ $\therefore \mathbf{x} = \begin{bmatrix} 2 \\ 3 \end{bmatrix}$

Any multiple is also an eigenvector, e.g., $\mathbf{x} = \begin{bmatrix} 1 \\ 3/2 \end{bmatrix}$ or $\mathbf{x} = \begin{bmatrix} 2/\sqrt{13} \\ 3/\sqrt{13} \end{bmatrix}$

1.41 **a)** $\dfrac{dy}{dx} = 6x^2 - 3 = 6(1)^2 - 3 = 3$

1.42 **c)** $\dfrac{dy}{dx} = \dfrac{1}{x} + e^x \cos x + e^x \sin x = 1 + e\cos 1 + e\sin 1 = 4.76.$ $(\cos 1 = \cos 57.3°)$

1.43 **d)** $\dfrac{dy}{dx} = 3x^2 - 3 = 0.$ $\therefore x^2 = 1.$ $\therefore x = \pm 1.$ $\dfrac{d^2y}{dx^2} = 6x.$ $\therefore x = -1$ is a maximum.

1.44 **c)** $y' = 3x^2 - 3.$ $y'' = 6x.$ $\therefore x = 0$ is inflection.

1.45 **a)** $\lim\limits_{x \to \infty} \dfrac{2x^2 - x}{x^2 + x} = \lim\limits_{x \to \infty} \dfrac{4x - 1}{2x + 1} = \lim\limits_{x \to \infty} \dfrac{4}{2} = 2$

1.46 **c)** $\eta(x + h) = \eta + h\eta' + \dfrac{h^2}{2}\eta''$

1.47 **b)** $e^x \sin x = \left(1 + x + \dfrac{x^2}{2}\right)\left(x - \dfrac{x^3}{6}\right) = x + x^2 + \dfrac{x^3}{2} - \dfrac{x^3}{6} = x + x^2 + x^3/3$

1.48 **d)** Area $= \int\limits_4^9 x\,dy = \int\limits_4^9 y^{1/2}\,dy = \dfrac{2}{3}(27 - 8) = 12\tfrac{2}{3}$

1.49 e) $\text{Area} = \int_0^4 (x_2 - x_1)\,dy = \int_0^4 \left(2y^{1/2} - \frac{y^2}{4}\right) dy$

$$= 2 \times \frac{2}{3} \times 8 - \frac{1}{12} \times 64 = 16/3$$

1.50 c) $V = \int_0^2 2\pi y\, x\, dy = 2\pi \int_0^2 y^3 dy = 2\pi \times \frac{2^4}{4} = 8\pi$

1.51 a) $\int_0^2 \left(e^x + \sin x\right) dx = e^x - \cos x\Big|_0^2 = e^2 - 1 - \cos 2 + 1 = 7.81$

1.52 a) $\int_0^1 e^x \sin x\, dx = e^x \sin x\Big|_0^1 - \int_0^1 e^x \cos x\, dx$

$$\underbrace{\begin{array}{ll} u = \sin x & dv = e^x dx \\ du = \cos x\, dx & v = e^x \end{array}}_{\text{1st integral}}$$

$$\underbrace{\begin{array}{ll} u = \cos x & dv = e^x dx \\ v = e^x & du = -\sin x\, dx \end{array}}_{\text{2nd integral}}$$

$\therefore \int_0^1 e^x \sin x\, dx = e\sin 1 - \left[e^x \cos x\Big|_0^1 + \int_0^1 e^x \sin x\, dx \right]$

$\therefore 2\int_0^1 e^x \sin x\, dx = e\sin 1 - e\cos 1 + 1 \times 1 = 1.819.$ $\therefore \int_0^1 e^x \sin x\, dx = 0.909$

1.53 e) $\int x \cos x\, dx = x\sin x - \int \sin x\, dx = x\sin x + \cos x + C$

$$\underbrace{\begin{array}{ll} u = x & dv = \cos x\, dx \\ du = dx & v = \sin x \end{array}}_{\text{1st integral}}$$

1.54 b) linear and nonhomogeneous. The term (+2) makes it nonhomogeneous.
$x^2 y'$ is linear.

1.55 a) $\frac{dy}{dx} = -2xy.$ $\frac{dy}{y} = -2x\,dx.$ $\therefore \ln y = -x^2 + C.$ $\ln 2 = 0 + C.$ $\therefore C = \ln 2.$

$y(2) = \exp\left(-2^2 + \ln 2\right) = 0.0366$

1.56 b) $\frac{dy}{dx} = -2x.$ $dy = -2x\,dx.$ $\therefore y = -x^2 + C.$ $1 = 0 + C.$ $\therefore C = 1.$

$y(10) = -10^2 + 1 = -99$

1.57 a) $2m^2 + m + 50 = 0.$ $\therefore m = \frac{-1 \pm \sqrt{1 - 400}}{4} = -\frac{1}{4} \pm 4.99i.$

$\therefore y(t) = e^{-1/4}\left(A\cos 4.99t + B\sin 4.99t\right).$

$\therefore \omega = 4.99 \text{ rad/s}.$ $\therefore f = \frac{\omega}{2\pi} = \frac{4.99}{2\pi} = 0.794$ hertz

1.58 a) $m^2 + 16 = 0.$ $\therefore m = \pm 4i.$ $\therefore y(t) = C_1 \cos 4t + C_2 \sin 4t$

1.59 **e)** $m^2 + 8m + 16 = 0.$ $(m+4)^2 = 0.$ $m = -4, -4.$ $\therefore y(t) = C_1 e^{-4t} + C_2 t e^{-4t}$

1.60 **d)** $m^2 - 5m + 6 = 0.$ $m = \dfrac{5 \pm \sqrt{25-24}}{2} = 3, 2.$ $\therefore y_h = C_1 e^{3t} + C_2 e^{2t}.$

Assume $y_p = Ae^t.$ Then $Ae^t - 5Ae^t + 6Ae^t = 4e^t.$ $\therefore A = 2.$

1.61 **a)** homogeneous: $m^2 + 16 = 0.$ $\therefore m = \pm 4i.$ $\therefore y_h(t) = C_1 \sin 4t + C_2 \cos 4t$

particular: $y_p = At \cos 4t.$ (This is resonance.) $\dot{y}_p = A\cos 4t - 4At \sin 4t$

$\therefore -4A \sin 4t - 4A \sin 4t - 16At \cos 4t + 16At \cos 4t = 8 \sin 4t$

$\therefore -8A = 8.$ $A = -1.$ $\therefore y = y_h + y_p = C_1 \sin 4t + C_2 \cos 4t - t \cos 4t$

1.62 **d)** $s^2 Y - 2s + 5sY - 10 + 6Y = 0.$ $Y = \dfrac{2s+10}{s^2 + 5s + 6} = \dfrac{-4}{s+3} + \dfrac{6}{s+2}.$

$\therefore y(t) = -4e^{-3t} + 6e^{-2t}$

1.63 **e)** $s^2 Y - 12 + Y = \dfrac{120}{(s+1)^2}.$ $Y = \dfrac{12}{s^2 + 1} + \dfrac{120}{(s+1)^2 (s^2 + 1)}$

$= \dfrac{12}{s^2 + 1} + \dfrac{60}{s+1} + \dfrac{60}{(s+1)^2} - \dfrac{60s}{s^2 + 1}$

$\therefore y(t) = 12 \sin t - 60 \cos t + 60\left(e^{-t} + t e^{-t}\right)$

1.64 **c)** $\mathbf{A} \cdot \mathbf{B} = 3 \cdot 10 + (-6) \cdot 4 + 2(-6) = -6$

1.65 **b)** $\mathbf{A} \times \mathbf{B} = (2\mathbf{i} - 5\mathbf{k}) \times \mathbf{j} = 2\mathbf{i} \times \mathbf{j} - 5\mathbf{k} \times \mathbf{j} = 2\mathbf{k} - 5(-\mathbf{i}) = 5\mathbf{i} + 2\mathbf{k}$

1.66 **e)** $\mathbf{i_B} = (6\mathbf{i} + 3\mathbf{j} - 2\mathbf{k}) \big/ \sqrt{6^2 + 3^2 + 2^2} = \dfrac{1}{7}(6\mathbf{i} + 3\mathbf{j} - 2\mathbf{k})$

$\mathbf{A} \cdot \mathbf{i_B} = \left[14 \cdot 6 - 7(3)\right]/7 = 12 - 3 = 9$

1.67 **a)** $\left[(x\mathbf{i} + y\mathbf{j} + z\mathbf{k}) - (2\mathbf{i} - 4\mathbf{j} + 6\mathbf{k})\right] \cdot (2\mathbf{i} - 4\mathbf{j} + 6\mathbf{k}) = 0.$

$2(x-2) - 4(y+4) + 6(z-6) = 0.$ $\therefore 2x - 4y + 6z = 56$

1.68 **b)** $|\mathbf{A} \times \mathbf{B}| = |15\mathbf{i} - 10\mathbf{j} - 24\mathbf{k}| = \sqrt{901} \cong 30$

1.69 **e)** $\nabla \phi = 2x\mathbf{i} + 6y\mathbf{j} - 3\mathbf{k} = 2\mathbf{i} + 6\mathbf{j} - 3\mathbf{k}.$ $|\nabla \phi| = 7.$ $\therefore \mathbf{i_n} = \nabla \phi / |\nabla \phi| = (2\mathbf{i} + 6\mathbf{j} - 3\mathbf{k})/7$

1.70 **d)** $\nabla \cdot u = 2x + 2y + 2z = 2 + 2 + 2 = 6$

1.71 **e)** $\nabla \times \mathbf{u} = (0-0)\mathbf{i} + (0-0)\mathbf{j} + (0-0)\mathbf{k} = 0$

1.72 **a)** $\nabla \cdot \mathbf{u} = 0 + 0 + 0 = 0.$ $\therefore$ solenoidal. $\nabla \times \mathbf{u} = (x-x)\mathbf{i} + (y-y)\mathbf{j} + (z-z)\mathbf{k} = 0.$

$\therefore$ conservative and solenoidal.

Statics

by George E. Mase

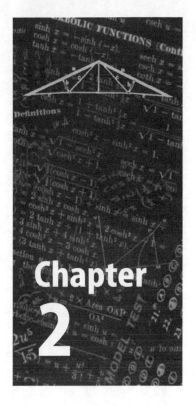

Chapter 2

Solutions to Practice Problems

2.1 a)
$$\mathbf{i}_B = \frac{\mathbf{i} - 2\mathbf{j} - 2\mathbf{k}}{\sqrt{1+4+4}} = \tfrac{1}{3}(\mathbf{i} - 2\mathbf{j} - 2\mathbf{k})$$

$$\mathbf{A} \cdot \mathbf{i}_B = (15\mathbf{i} - 9\mathbf{j} + 15\mathbf{k}) \cdot \tfrac{1}{3}(\mathbf{i} - 2\mathbf{j} - 2\mathbf{k}) = 5 + 6 - 10 = 1$$

2.2 c)
$$\mathbf{A} + \mathbf{B} + \mathbf{C} = (2\mathbf{i} + 5\mathbf{j}) + (6\mathbf{i} - 7\mathbf{k}) + (2\mathbf{i} - 6\mathbf{j} + 10\mathbf{k}) = 10\mathbf{i} - \mathbf{j} + 3\mathbf{k}$$

$$\text{magnitude} = \sqrt{10^2 + 1^2 + 3^2} = 10.49$$

2.3 e)
$$\mathbf{M} = \mathbf{r} \times \mathbf{F} = (4\mathbf{i} - 6\mathbf{j} + 4\mathbf{k}) \times (200\mathbf{i} + 400\mathbf{j}). \quad M_y = 4 \times 200 = 800 \quad \text{since } \mathbf{k} \times \mathbf{i} = \mathbf{j}$$

2.4 c)
$$\mathbf{M} = \mathbf{r}_1 \times \mathbf{F}_1 + \mathbf{r}_2 \times \mathbf{F}_2 = (2\mathbf{i} - 4\mathbf{k}) \times (50\mathbf{i} - 40\mathbf{k}) + (-4\mathbf{i} + 2\mathbf{j}) \times (60\mathbf{j} + 80\mathbf{k})$$

$$M_x = 2 \times 80 = 160 \quad \text{since } \mathbf{j} \times \mathbf{k} = \mathbf{i}$$

2.5 e) Concurrent $\Rightarrow$ all pass through a point.

Coplanar $\Rightarrow$ all in the same plane.

The forces are three-dimensional.

2.6 **b)** $\sum \mathbf{F} = 0.$ $\therefore \mathbf{R} + 141\mathbf{i} - 141\mathbf{j} - 200\mathbf{i} - 100\mathbf{k} = 0$
$\therefore \mathbf{R} = 59\mathbf{i} + 141\mathbf{j} + 100\mathbf{k}$

2.7 **c)** $\sum \mathbf{M} = 0.$ $\therefore \mathbf{M}_A + (4\mathbf{j} - 3\mathbf{k}) \times (-100\mathbf{k}) - 3\mathbf{k} \times (-200\mathbf{i}) + 4\mathbf{i} \times (141\mathbf{i} - 141\mathbf{j}) = 0$
$\therefore \mathbf{M}_A = 400\mathbf{i} - 600\mathbf{j} + 564\mathbf{k}$

2.8 **b)**   They must be concurrent, otherwise a resultant moment would occur.

2.9 **a)**   It is a two-force body.

2.10 **b)** $\sum M_A = 0.$ $F_B \times 8 = 400 \times 4 + 400 \times 6.$ $\therefore F_B = 500 \text{ N}$

2.11 **a)** $M_A = 400 \times 8 + 400 \times 6 = 5600 \text{ N} \cdot \text{m}$

2.12 **b)** $\sum M_B = 0.$ $6F_A = 4 \times 300 + 600 \times 3/2.$ $\therefore F_A = 350 \text{ N}$

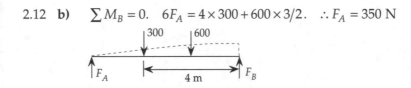

2.13 **c)** $M_A = 0.6 \times 100 - 141 \times 0.6 + 141 \times 0.8 = 88.2$

2.14 **a)** $M = 100 \sin 45° \times 4 = 282.8 \text{ cw}$

2.15 **c)** $\sum M_A = 0.$ $\therefore 6 \times 70.7 = 2 \times 0.866 F_1.$ $\therefore F_1 = 245$
$\sum F_x = 0.$ $\therefore -70.7 - 245 \times 0.5 + F_{Ax} = 0.$ $\therefore F_{Ax} = 193$
$\sum F_y = 0.$ $\therefore -70.7 - 245 \times 0.866 + F_{Ay} = 0.$ $\therefore F_{Ay} = 283$
$$\therefore F_A = \sqrt{F_{Ax}^2 + F_{Ay}^2} = \sqrt{193^2 + 283^2} = 343$$

2.16 **e)** $\sum M_A = 0.$ $\therefore 2F_B + 1.2 \times 200 - 141.4 \times 2 - 141.4 \times 1.2 + 50 = 0.$ $\therefore F_B = 81.2$
$\sum F_x = 0.$ $\therefore F_{Ax} - 200 + 141.4 = 0.$ $\therefore F_{Ax} = 58.6$
$\sum F_y = 0.$ $\therefore F_{Ay} + 81.2 - 141.4 = 0.$ $\therefore F_{Ay} = 60.2$
$$\therefore F_A = \sqrt{F_{Ax}^2 + F_{Ay}^2} = \sqrt{58.6^2 + 60.2^2} = 84.0$$

2.17 **e)** $\sum M_A = 0.$ $\therefore 500\ell + 200 \times 0.866\ell - F_C \times 2\ell = 0.$ $\therefore F_C = 337$
$0.866 F_{DC} = 337.$ $\therefore F_{DC} = 389$
$0.866 \times 389 = 0.866 F_{BD}.$ $\therefore F_{BD} = 389$
$-F_{DE} + 200 - 389 \times 0.5 - 389 \times 0.5 = 0.$ $\therefore F_{DE} = -189$

2.18 d) $\sum M_A = 0.$ $\therefore 5 \times 5000 = 10 \times F_C.$ $\therefore F_C = 2500$ $\therefore F_{DC} = 2500$

$0.707 F_{BD} = 2500.$ $\therefore F_{BD} = 3536$

$0.707 \times 3536 = F_{DE}.$ $\therefore F_{DE} = 2500$

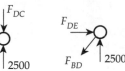

2.19 b) $\sum M_A = 0.$ $\therefore 4 \times 2000 + 6 \times 1000 = 8 F_C.$ $\therefore F_C = 1750$

$0.707 F_{DC} = 1750.$ $\therefore F_{DC} = 2475$

Sum forces in dir. of F_{DE} : $F_{DE} - 2475 + 1000 \times 0.707 = 0.$

$$\therefore F_{DE} = 1768$$

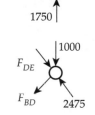

2.20 a) Sum forces in dir. of F_{FB} at F. $F_{FB} = 0.$

2.21 c) $\sum M_B = 0.$ $\therefore 12 F_F - 3 \times 4000.$ $\therefore F_F = 1000 \downarrow$

$\sum F_y = 0.$ $\therefore 0.8 F_{IC} = 1000.$ $\therefore F_{IC} = 1250$

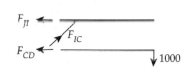

2.22 d) Cut vertically through link KA. Then $F_{KA} = 5000.$
Obviously, $F_{AL} = 0.$ $\therefore F_{AB} = 3000.$ $\therefore F_{BC} = 3000$

2.23 e) At E we see that $F_{EC} = 0.$ $\therefore$ At C, $F_{FC} = 0$

2.24 e) $9^2 = 6^2 + 5^2 - 2 \times 5 \times 6 \cos\theta.$ $\therefore \theta = 109.5°$

$6^2 = 9^2 + 5^2 - 2 \times 9 \times 5 \cos\alpha.$ $\therefore \alpha = 38.9°$

From pts E, C, F, B we see that $F_{EC} = F_{FC} = F_{FB} = F_{GB} = 0.$

Also, $F_A = F_{BC}.$ $\sum M_G = 0.$

$\therefore 5 \times F_A \sin 38.9° + 5000 \times 6 \sin 70.5° = 5000 \times 6 \cos 70.5°.$

$\therefore F_A = -5817$

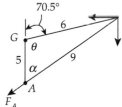

2.25 b)   Recognize that link BC is a two-force member. $\sum M_A = 0.$

$\therefore 0.2 \times 1000 + 100 = 0.08 \times F_{BC} \times 0.8 + 0.2 \times F_{BC} \times 0.6$

$\therefore F_{BC} = 1630.$ $A_x = 1630 \times 0.8 = 1304.$ $A_y = 1630 \times 0.6 - 1000 = -22$

$\therefore F_A = \sqrt{1304^2 + 22^2} = 1304$ N

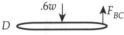

2.26 d) $1800 \times 0.8 = 0.6 F_{BC}.$ $\therefore F_{BC} = 2400$

$0.3 \times 0.6w = 0.6 \times 2400.$ $\therefore w = 8000$

2.27 d) $\sum M_A = 0.$ $\therefore 1.2 F_E = 0.8 \times 2400.$ $\therefore F_E = 1600$

$\therefore A_x = 2400.$ $A_y = 1600$

$\therefore F_A = \sqrt{2400^2 + 1600^2} = 2884$ N

2.28 a)   Link BD is a two-force member. $\therefore$ the force acts from D to B. Hence, the angles are found.

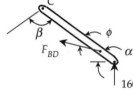

$120^2 = 100^2 + 100^2 - 2 \times 100 \times 100 \cos \beta.$ $\therefore \beta = 73.7°$

$\overline{BD}^2 = 60^2 + 40^2 - 2 \times 60 \times 40 \cos 73.7°.$ $\therefore BD = 62.1$

$\dfrac{62.1}{\sin 73.7°} = \dfrac{40}{\sin \phi}.$ $\therefore \phi = 38.2°.$ $\alpha = (180 - 73.7)/2 = 53.2°$

$\sum M_C = 0.$ $1600 \times 100 \cos 53.2° = 60 \times F_{BD} \sin 38.2°.$ $\therefore F_{BD} = 2587$

2.29 e) $\sum F_y = 0.$ $N \times 0.866 - 980 - 0.4N \times 0.5 = 0.$ $\therefore N = 1471$

$\sum F_x = 0.$ $F = 1471 \times 0.5 + 0.4 \times 1471 \times 0.866 = 1245$

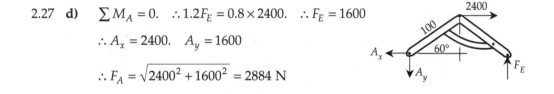

2.30 b) $N_1 = 490.$ $N_2 = 980.$ $\therefore F = 0.2(490 + 980) = 294$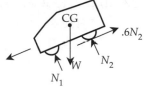

2.31 c) $\sum M_{\text{front wheel}} = 0.$ $\therefore 400 N_2 - W\cos\theta \times 200 + W\sin\theta \times 50 = 0$

$\sum F_x = 0.$ $\therefore 0.6 N_2 = W\sin\theta.$

$\therefore 400\left(W\sin\theta\right)/0.6 + 50 W\sin\theta = 200 W\cos\theta$

$\therefore \dfrac{\sin\theta}{\cos\theta} = \dfrac{200}{716.7} = \tan\theta.$ $\therefore \theta = 15.6^\circ$

2.32 b)   If $h < h_{\min}$ then sliding occurs, and $F_f = 0.4 N$.
 If $h > h_{\min}$ tipping occurs and $F_f < 0.4 N$.
 When $h = h_{\min}$, $F_f = 0.4 N = 0.4 W = F$.
 $\sum M_A = 0.$ $\therefore 4 W = hF = h \times 0.4 W.$ $\therefore h = 10$ cm

2.33 e) $\sum F_x = 0.$ $\therefore N_2 = 0.4 N_1.$ Also, $W = 980$
 $\sum M_A = 0.$ $\therefore W \cdot r = \left(N_1 + 0.4 N_1 + 2 \times 0.4 N_2\right) r.$
 $\therefore N_1 = 0.5814 W = 570$
 $\sum F_y = 0.$ $\therefore F = 980 - 570 - 0.16 \times 570 = 319$

2.34 b) $\sum F_x = 0.$ $\therefore N_2 = 0.4 N_1.$ $\sum F_y = 0.$ $\therefore N_1 + 0.4 N_2 = W.$
 $\therefore N_2 = 0.345 W$

$\sum M_A = 0.$ $\therefore \dfrac{L}{2} \times W\cos\theta = N_2 \times L\sin\theta + 0.4 N_2 \times L\cos\theta.$

This gives $\tan\theta = 1.049.$ $\therefore \theta = 46.4^\circ$

2.35 a) $F_B = F_D e^{-\mu\theta} = 800 e^{-0.5\pi} = 166\,\text{N}$

2.36 **a)** $\sum M_A = 0.$ $\therefore 200 \times 0.6 = 0.1 \times T_1 + 0.1 \times T_2.$

$T_1 = T_2 e^{0.4 \times 3\pi/2} = 6.59 T_2.$ Thus, $T_2 = 158$ and $T_1 = 1042.$

$\sum M_{\text{center}} = 0.$ $\therefore M = 0.1 \times (1042 - 158) = 88.4 \text{ N·m}$

2.37 **d)**   Let h = long end. m = mass/unit length. Then,
$(12 - 1.88 - h) m g e^{0.5\pi} = hmg.$ $\therefore h = 8.38 \text{ m}$

2.38 **d)** $\bar{x} = \dfrac{\displaystyle\int_0^3 xy\,dx}{\displaystyle\int_0^3 y\,dx} = \dfrac{\displaystyle\int_0^3 x^3\,dx}{\displaystyle\int_0^3 x^2\,dx} = \dfrac{3^4/4}{3^3/3} = 2.25$

2.39 **a)** $\bar{y} = \dfrac{\displaystyle\int_0^3 \frac{y}{2} y\,dx}{\displaystyle\int_0^3 y\,dx} = \dfrac{\frac{1}{2}\displaystyle\int_0^3 x^4\,dx}{\displaystyle\int_0^3 x^2\,dx} = \dfrac{3^5/10}{3^3/3} = 2.7$

2.40 **c)** $\bar{x} = \dfrac{\displaystyle\int_0^1 \left(\sqrt{x} - x^2\right) x\,dx}{\displaystyle\int_0^1 \left(\sqrt{x} - x^2\right) dx} = \dfrac{\frac{1}{5/2} - \frac{1}{4}}{\frac{1}{3/2} - \frac{1}{3}} = 0.45$

2.41 **b)** $\bar{y} = \dfrac{24 \times 3 + 6 \times 5}{6 \times 4 + 4 \times 3/2} = 3.4$

2.42 **e)** $\bar{y} = \dfrac{48 \times 3 + 12 \times 7 - \pi \times 6}{8 \times 6 + 3 \times 4 - \pi} = 3.68$

2.43 **b)** $\bar{x} = \dfrac{10 \times \frac{1}{2} + 5 \times 3.5 + 3 \times 7}{10 + 5 + 3} = 2.42$

2.44 **b)** $I_x = \int_0^3 y^3\,dx/3 = \int_0^3 x^6\,dx/3 = 3^7/21 = 104.1.$

With a horizontal strip: $I_x = \int_0^9 y^2(3-x)dy = \int_0^9 y^2\left(3-\sqrt{y}\right)dy = 9^3 - \dfrac{9^{7/2}}{7/2} = 104.1$

2.45 **e)** $I_x = 8\times 6^3/3 + \left(8\times 3^3/36 + 12\times 7^2\right) - \left(\pi\times 1^4/4 + \pi\times 6^2\right) = 1056$

2.46 **a)** $I_y = 12\times 12^3/3 - \left(8\times 8^3/12 + 64\times 6^2\right) = 4267.$ Or, alternatively:

$I_y = 8\times 2^3/3 + 4\times 12^3/3 + 8\times 2^3/12 + 16\times 11^2 = 4267$

2.47 **a)** $I_{\text{edge}} = I_{\text{c.g.}} + Md^2 = \dfrac{1}{12}M\left(b^2 + b^2\right) + M\dfrac{b^2}{2} = \dfrac{2}{3}Mb^2$

2.48 **e)** $I_x = \dfrac{1}{3}(6m)\times 6^2 \times 2 + 8m\times 6^2 = 432$ with $m = 1$

Mechanics of Materials

by George E. Mase

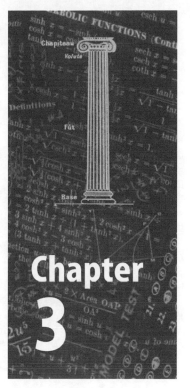

Chapter

3

Solutions to Practice Problems

Metric Units

3.1 **b)** Isotropic

3.2 **e)** Poisson's Ratio

3.3 **d)**

3.4 **d)** $\sigma = E\delta/L = P/\pi r^2$

$\therefore P = E\,\delta\pi r^2/L = 210 \times 10^9 \times 0.001 \times \pi \times 0.01^2/1 = 66\,000$ N

3.5 **d)** $\dfrac{P}{\pi r^2} = E\dfrac{\delta}{L}. \quad \therefore \delta = \dfrac{14\,000 \times 30}{\pi \times 0.01^2 \times 210 \times 10^9} = 0.0064$ m

3.6 **a)** $\dfrac{P}{A} = E\dfrac{\delta}{L}. \quad \therefore P = \dfrac{AE\delta}{L} = 0.025 \times 0.003 \times \left(210 \times 10^9\right) \times \dfrac{0.0008}{1} = 12\,600$ N

3.7 **b)** $\dfrac{P}{A} = G\dfrac{\delta}{L}. \quad \therefore G = \dfrac{PL}{A\delta} = \dfrac{20\,000 \times 0.15}{0.012 \times 0.2 \times 0.00003} = 41.7 \times 10^9$ Pa

3.8 **a)** $P/A = E\varepsilon. \quad \varepsilon = \dfrac{600\,000}{\pi \times 0.025^2 \times 210 \times 10^9} = 0.001455$

$\Delta d = v\varepsilon d = 0.28 \times 0.001455 \times 5 = 0.00204$ cm

$d = d - \Delta d = 5 - 0.00204 = 4.9980$ cm

3.9   **b)** $P/A = E\varepsilon$. $\varepsilon = \dfrac{1\,500\,000}{\pi \times 0.06^2 \times 70 \times 10^9} = 0.00190$

$\Delta d = \nu\varepsilon d = 0.33 \times 0.0019 \times 12.015 = 0.0075$ cm

$d = d - \Delta d = 12.015 - 0.0075 = 12.008$ cm

3.10 **e)** $\sigma = E\delta/L$. $100 \times 10^6 = 210 \times 10^9 (\delta/L)$. $\therefore$ $\delta = 4.76 \times 10^{-4} L$

$\delta_T = \alpha L \Delta T = 11.7 \times 10^{-6} \times 30L = 3.51 \times 10^{-4} L$. $\delta_{final} = \delta - \delta_T = 1.25 \times 10^{-4} L$

$\therefore \sigma = \left(210 \times 10^9\right) \times \left(1.25 \times 10^{-4}\right) = 26.2 \times 10^6$ Pa

3.11 **a)** $\delta = \alpha L \Delta T = \left(11.7 \times 10^{-6}\right) \times 300 \times 75 = 0.263$ m

3.12 **d)** $\dfrac{\delta}{L} = \dfrac{\sigma}{E} = \alpha \Delta T$. $\dfrac{70 \times 10^6}{70 \times 10^9} = 23 \times 10^{-6}(30 - T)$. $\therefore T = -13.5°C$

3.13 **d)**   It expands at a different rate.

3.14 **a)** $\tau = \dfrac{Tr}{J} = \dfrac{200 \times 0.03}{\pi \times 0.06^4/32} = 4.72 \times 10^6$ Pa

3.15 **e)** $J = \pi\left(a_1^4 - a_2^4\right)/2 = \pi\left(0.03^4 - .025^4\right)/2 = 65.9 \times 10^{-8}$

$\tau = \dfrac{Tr}{J} = \dfrac{200 \times 0.03}{65.9 \times 10^{-8}} = 9.10 \times 10^6$ Pa

3.16 **a)** $T = \dfrac{\tau J}{r} = \dfrac{\left(140 \times 10^6\right) \times \pi \times .05^4/2}{0.05} = 27\,500$ N·m

3.17 **d)** $\theta = \dfrac{TL}{JG} = \dfrac{200 \times 0.3 \times 0.5}{\left(83 \times 10^9\right) \times \pi \times .05^4/2} = 0.368$ rad or 21.1°

3.18 **d)** $\tau_{max} = \dfrac{Tr}{J} = \dfrac{1200 \times 0.04}{\pi \times 0.04^4/2} = 11.9 \times 10^6$ Pa.

$\therefore \sigma_{max} = 11.9 \times 10^6$ Pa.

3.19 **b)** $\sigma = My/I$. σ_{max} occurs at $y = y_{max}$

3.20 **a)**   A triangle.

3.21 **a)** $\sum M_{right} = 0$. $\therefore 8F_{left} = 4\,000 \times 4 + 4\,000 \times 2$. $\therefore F = 3\,000$.

$M_A = 3\,000 \times 4 = 12\,000$ N·m

3.22 **b)** $M_A = 4\,000 \times 4 + 4\,000 \times 2 = 24\,000$ N·m

3.23 **b)** $4F_{right} = 24\,000 \times 2 + 4\,000 \times 6.$ $\therefore F_{right} = 18\,000$ N. $4F_{left} = 24\,000 \times 2 - 4\,000 \times 2.$

$\therefore F_{left} = 10\,000$ N. M_{max} = area under diagram = $10\,000 \times 1.667/2 = 8\,330$ N·m.

$\sigma = \dfrac{My}{I} = \dfrac{8\,330 \times 0.1}{971 \times 10^{-8}} = 85.8 \times 10^6$ Pa

3.24 **c)** Compression occurs in bottom fibers over right support.

There, $M = 4\,000 \times 2 = 8\,000$ N·m. $\sigma = \dfrac{My}{I} = \dfrac{8\,000 \times 0.1}{971 \times 10^{-8}} = 82.4 \times 10^6$ Pa

3.25 **e)** $\tau_{max} = \dfrac{VQ}{Ib} = \dfrac{14\,000(0.002 \times 0.05)}{\left(971 \times 10^{-8}\right) \times 0.02} = 7.21 \times 10^6$ Pa

3.26 **c)** τ_{max} occurs on the N.A. with a sudden decrease when b goes from 2 to 16 cm.

Also, it is a parabolic distribution.

3.27 **c)** $\sigma = \dfrac{My}{I}.$ $\dfrac{I}{y} = \dfrac{M}{\sigma} = \dfrac{24\,000}{140 \times 10^6} = 171 \times 10^{-6}$ m^3

3.28 **b)** Using the area under the curve: $M_{max} = 1000 \times 3 + 3000 \times 3/2 = 7500$ N·m

$\sigma_{max} = \dfrac{My}{I} = \dfrac{7500 \times 0.025}{0.1 \times 0.05^3/12} = 180 \times 10^6$ Pa

3.29 **e)** $\sigma_{max} = \dfrac{My}{I} = \dfrac{7500 \times 0.05}{0.05 \times 0.1^3/12} = 90 \times 10^6$ Pa

3.30 **a)** $V_{max} = 1000.$ $\tau_{max} = \dfrac{VQ}{Ib} = \dfrac{1000(0.025 \times 0.05 \times 0.0125)}{\left(0.05 \times 0.05^3/12\right) \times 0.05} = 600 \times 10^3$ Pa

3.31 **d)** $\delta = \dfrac{PL^3}{48EI} = \dfrac{2000 \times 6^3}{48 \times \left(210 \times 10^9\right) \times 0.05^4/12} = 0.0823$ m

3.32 **e)** $\delta = \dfrac{PL^3}{48EI} + \dfrac{5wL^4}{384EI}.$ $I = \dfrac{bh^3}{12} = \dfrac{0.1 \times 0.05^3}{12} = 1.04 \times 10^{-6}$

$= \dfrac{2000 \times 6^3}{48 \times \left(210 \times 10^9\right) \times 1.04 \times 10^{-6}} + \dfrac{1000 \times 5 \times 6^4}{384 \times \left(210 \times 10^9\right) \times 1.04 \times 10^{-6}} = 0.118$ m

3.33 **a)** $\delta = \theta L_2 = \dfrac{PL^2}{16EI} \times 4.$ $0.1 = \dfrac{P \times 8^2 \times 4}{16 \times \left(210 \times 10^9\right)\pi \times .025^4/4}.$ $\therefore P = 403$ N

3.34 **d)** $\tau_{max} = \dfrac{1}{2}\sqrt{\left(\sigma_x - \sigma_y\right)^2 + 4\tau^2} = \dfrac{1}{2}\sqrt{60^2 + 4 \times 40^2} = 50$ MPa

3.35 **a)** $\alpha_{max} = \dfrac{1}{2}\left(\sigma_x + \sigma_y\right) + \tau_{max} = 0 + 40 = 40$ MPa

3.36 **c)** $\tau_{\max} = \frac{1}{2}\sqrt{(30+50)^2 + 4 \times 30^2} = 50$ MPa

3.37 **d)** $\tau = Tr/J = 600 \times 0.025 \Big/ \dfrac{\pi \times .025^4}{2} = 24.45 \times 10^6$ Pa

$\sigma = P/A = 40\,000 \Big/ \pi \times 0.025^2 = 20.37 \times 10^6$ Pa

$\therefore \tau_{\max} = \frac{1}{2}\sqrt{20.37^2 + 4 \times 24.45^2} = 26.5 \times 10^6$ Pa

3.38 **c)** $\sigma_{\max} = \left(\frac{1}{2} \times 20.37 + 26.5\right) \times 10^6 = 36.7 \times 10^6$ Pa

3.39 **b)** $My/I = 3\,200 \times 0.025 \Big/ \dfrac{\pi \times .025^4}{4} = 261 \times 10^6$ comp.

$P/A = 40\,000 \Big/ \pi \times 0.025^2 = 20.4 \times 10^6$ Pa tension.

$\sigma_A = (261 - 20.4) \times 10^6 = 241 \times 10^6$ Pa

3.40 **c)** $\tau_{\max} = \sigma/2 = 120 \times 10^6$ Pa. $VQ/Ib = 0$ on outer fibers.

3.41 **c)** $\tau = \dfrac{Tr}{J} = \dfrac{8\,000 \times 0.25 \times 0.025}{\pi \times .025^4 / 2} = 81.5 \times 10^6$ Pa

$\sigma = \dfrac{My}{I} = \dfrac{8\,000 \times 0.4 \times 0.025}{\pi \times .025^4 / 4} = 261 \times 10^6$ Pa

$\tau_{\max} = \frac{1}{2}\sqrt{261^2 + 4 \times 81.5^2} \times 10^6 = 154 \times 10^6$ Pa

3.42 **a)** $\sigma_{\max} = \left(\frac{1}{2} \times 261 + 154\right) \times 10^6 = 284 \times 10^6$ Pa

3.43 **c)** $M_{\max} = FL_2 = 2000 \times 1 = 2000$ N·m.

$\sigma = \dfrac{My}{I} = \dfrac{2000 \times 0.025}{0.02 \times 0.05^3 / 12} = 240 \times 10^6$ Pa.

3.44   **d)** $\dfrac{VQ}{Ib} = \dfrac{2000 \times (0.02 \times 0.025) \times 0.0125}{\left(0.02 \times 0.05^3/12\right) \times 0.02} = 3 \times 10^6$ Pa.

3.45   **a)**   The force F provides a torque of 2000 N·m and a force of 2000 N acting on the end of the steel shaft. This is best described by (a) since the top fibers experience tension. There is no normal stress in the circumferential direction, ruling out (b), (c) and (d).

3.46   **e)** $\sigma = \dfrac{My}{I} = \dfrac{(2000 \times 2) \times 0.025}{\pi \times 0.025^4/4} = 326 \times 10^6$ Pa.

$\tau = \dfrac{Tr}{J} = \dfrac{(2000 \times 1) \times 0.025}{\pi \times 0.025^4/2} = 81.5 \times 10^6$ Pa.

$\therefore \sigma_{max} = 163 + \left[163^2 + 81.5^2\right]^{1/2}$

$\qquad = 345$ MPa.

3.47   **d)** $\delta = \dfrac{PL^3}{3EI} = \dfrac{2000 \times 2^3}{3 \times 210 \times 10^9 \times \pi \times 0.025^4/4} = 0.0828$ m.

3.48   **b)** $\sigma_t = pD/2t.$ $\therefore p = 180 \times 10^6 \times 2 \times 0.005/0.8 = 2250 \times 10^3$ Pa

3.49   **c)** $\sigma_u = pD/4t$ $\therefore t = \dfrac{8000 \times 10^3 \times 1.2}{4 \times 200 \times 10^6} = 0.012$ m

3.50   **e)** $\tau_{max} = \dfrac{1}{2}\sqrt{(200 - 200) + 0 \times 4} = 0$

3.51   **a)** $\left(\dfrac{\Delta L}{L}\right)_s = \left(\dfrac{\Delta L}{L}\right)_c.$ $\therefore \varepsilon_s = \varepsilon_c.$ $\therefore \sigma_s = \dfrac{E_s}{E_c}\sigma_c = 10.5\,\sigma_c.$

$F_s + F_c = 2\,000\,000$ or $A_s\sigma_s + A_c\sigma_c = 2\,000\,000.$

$\sigma_s\left[\pi\left(0.137^2 - 0.125^2\right) + \pi \times \dfrac{0.125^2}{10.5}\right] = 2 \times 10^6.$ $\therefore \sigma_s = 137 \times 10^6$ Pa.

3.52   **d)** $n = E/E_{min} = 3.$ The area is transformed :

$I_t = \dfrac{0.12 \times 0.24^3}{12} - \dfrac{0.03 \times 0.18^3}{12} = 1.237 \times 10^{-4}$ m^4.

$\therefore \sigma_{al} = \dfrac{My}{I} = \dfrac{2000 \times 2 \times 0.12}{1.237 \times 10^{-4}} = 3.88 \times 10^6$ Pa , $\sigma_s = \dfrac{nMy}{I} = \dfrac{3 \times 2000 \times 2 \times 0.09}{1.237 \times 10^{-4}} = 4.37 \times 10^6$ Pa

3.53   **c)** $n = E/E_{min} = 3.$ The area is transformed :

$$I_t = \frac{0.36 \times 0.24^3}{12} - \frac{0.33 \times 0.18^3}{12} = 2.54 \times 10^{-4} \ m^4.$$

$$\therefore \sigma_s = \frac{nMy}{I} = \frac{3 \times 4000 \times 0.12}{2.54 \times 10^{-4}} = 5.67 \times 10^6 \ Pa$$

3.54   **d)** $60 = \dfrac{L}{r} = \dfrac{L}{\sqrt{I/A}} = \dfrac{L}{\sqrt{0.1 \times \left(0.1^3/12\right)/0.01}}.$ $\therefore L = 1.73 \ m$

3.55   **c)**   Assume $P = 4000$ N using a factor of safety of 2.

$$4000 = \frac{\pi^2 \times 70 \times 10^9 \times \pi \times \left(0.1^4/64\right)}{4L^2}. \quad \therefore L = 14.55 \ m$$

3.56   **e)** $P_{cr} = 4\pi^2 EI/L^2 = \alpha \Delta TEA. \quad \Delta T = \dfrac{4\pi^2 I}{\alpha A L^2} = \dfrac{4\pi^2 \times \pi \times \left(0.02^4/64\right)}{11.7 \times 10^{-6} \times \pi \times 0.01^2 \times 4^2} = 5.27 \ ^\circ C$

3.57   **b)** $P_{cr} = 4\pi^2 EI/L^2 = 30\,000. \quad \therefore \pi^2 EI/L^2 = 7500. \quad \therefore P_{cr} = \pi^2 EI/4L^2 = 7500/4 = 1875 \ N$

Solutions to Practice Problems

English Units

3.1   **b)**   Isotropic

3.2   **e)**   Poisson's Ratio

3.3   **d)**

3.4   **d)** $\sigma = E\delta/L = P/\pi r^2. \quad \therefore P = E\delta \pi r^2/L = 30 \times 10^6 \times 0.04 \times \pi \left(1/4\right)^2/48 = 4909$ lb.

3.5   **d)** $\dfrac{P}{\pi r^2} = E\dfrac{\delta}{L}. \quad \therefore \delta = \dfrac{PL}{\pi r^2 E} = \dfrac{3500 \times 100 \times 12}{\pi \times \left(1/4^2\right) \times 30 \times 10^6} = 0.71$ in

3.6   **a)** $\dfrac{P}{A} = E\dfrac{\delta}{L}. \quad \therefore P = \dfrac{AE\delta}{L} = \dfrac{1}{8} \times 1 \times 30 \times 10^6 \times \dfrac{1/32}{36} = 3255$ lb.

3.7   **b)** $\dfrac{P}{A} = G\dfrac{\delta}{L}. \quad \therefore G = \dfrac{PL}{A\delta} = \dfrac{5000 \times 6}{8 \times \left(1/2\right) \times 0.0012} = 6.25 \times 10^6$

3.8 **a)** $\dfrac{P}{A} = E\varepsilon. \quad \therefore \varepsilon = \dfrac{150,000}{\pi \times 1^2 \times 30 \times 10^6} = 0.00159$

$\therefore \Delta d = v\varepsilon d = 0.28 \times 0.00159 \times 2 = 0.00089. \quad \therefore d_{after} = 2 - 0.00095 = 1.9991 \text{ in}$

3.9 **b)** $\dfrac{P}{A} = E\varepsilon. \quad \therefore \varepsilon = \dfrac{400,000}{\pi \times (5.923/2)^2 \times 10 \times 10^6} = 0.00145$

$\Delta d = v\varepsilon d = 0.33 \times 0.00145 \times 5.923 = 0.00283.$

$\therefore d_f = d - \Delta d = 5.923 - 0.00283 = 5.920 \text{ in}$

3.10 **e)** $\sigma = E\delta/L. \quad 16,000 = 30 \times 10^6 (\delta/L). \quad \therefore \delta = 5.33 \times 10^{-4} L.$

$\delta_T = \alpha L \Delta T = 6.5 \times 10^{-6} \times 50L = 3.25 \times 10^{-4} L.$

$\therefore \delta_{final} = 2.08 \times 10^{-4} L. \quad \therefore \sigma = 30 \times 10^6 \times 2.08 \times 10^{-4} = 6,240 \text{ psi}$

3.11 **a)** $\delta = \alpha L \Delta T = 6.5 \times 10^{-6} \times 1000 \times 12 \times 130 = 10.1"$

3.12 **d)** $\dfrac{\delta}{L} = \dfrac{\sigma}{E} = \alpha \Delta T. \quad \therefore \dfrac{10,000}{10 \times 10^6} = 12.8 \times 10^{-6}(80 - T). \quad \therefore T = 1.88°\text{F}$

3.13 **d)** It expands at a different rate.

3.14 **a)** $\tau = \dfrac{Tr}{J} = \dfrac{2000 \times 1}{\pi \times 1^4/2} = 1,273 \text{ psi}$

3.15 **e)** $J = \pi\left(a_1^4 - a_2^4\right)/2 = \pi\left(1^4 - .875^4\right)/2 = 0.650 \text{ in}^4$

$\tau = \dfrac{Tr}{J} = 2,000 \times 1/0.650 = 3,077 \text{ psi}$

3.16 **a)** $T = \dfrac{\tau J}{r} = 20,000 \times \left(\pi 2^4/2\right)/2 = 251,300 \text{ in} \cdot \text{lb or } 20,940 \text{ ft} \cdot \text{lb}$

3.17 **d)** $\theta = \dfrac{TL}{JG} = \dfrac{160 \times 24 \times 18}{\left[\pi(7/16)^4/2\right] \times 12 \times 10^6} = 0.1001 \text{ rad } (5.73°)$

3.18 **d)** $\tau_{max} = \dfrac{Tr}{J} = \dfrac{(1200 \times 12) \times 2}{\pi \times 2^4/2} = 1146 \text{ psi}. \quad \therefore \sigma_{max} = 1146 \text{ psi}.$

3.19 **b)** $\sigma = My/I. \quad \therefore \sigma = \sigma_{max} \text{ at } y = y_{max}$

3.20 **a)** A triangle.

3.21 **a)** $\sum M_{right\,end} = 0. \quad \therefore 20F_{left} = 1,000 \times 10 + 1,000 \times 5. \quad \therefore F_{left} = 750.$

$M_A = 750 \times 10 = 7,500 \text{ lb} \cdot \text{ft}$

3.22 **b)** $M_A = 1,000 \times 10 + 1,000 \times 5 = 15,000 \text{ ft} \cdot \text{lb}$

3.23 **a)** $10F_{right} = 5,000 \times 5 + 1,000 \times 14. \quad \therefore F_{right} = 3,900$ lb.

$10F_{left} = 5,000 \times 5 - 1,000 \times 4. \quad \therefore F_{left} = 2,100$ lb.

$M_{max} = 2,100 \times (4.2/2) = 4,410 \text{ ft} \cdot \text{lb} = \text{area under V-diagram.}$

$\sigma = \dfrac{My}{I} = \dfrac{4,410 \times 12 \times 5}{60.67} = 4,361$ psi

3.24 **b)** Compression occurs in bottom fibers over right support.

$\sigma = \dfrac{My}{I} = \dfrac{4,000 \times 12 \times 5}{60.67} = 3,956$ psi

3.25 **e)** $\tau_{max} = \dfrac{VQ}{Ib} = \dfrac{2,900 \times (5 \times 2.5)}{60.67 \times 1} = 597$ psi

3.26 **c)** τ_{max} occurs on the N.A. with a sudden decrease when b goes from 1" to 8".

Also, it is parabolic.

3.27 **c)** $\sigma = \dfrac{My}{I}. \quad \therefore \dfrac{I}{y} = \dfrac{M}{\sigma} = \dfrac{15,000 \times 12}{2,000} = 9 \text{ in}^3$

3.28 **b)** Area under curve: $M_{max} = 250 \times 8 + 800 \times (8/2) = 5,200$

$\sigma = \dfrac{My}{I} = \dfrac{5,200 \times 12 \times 1}{4 \times 2^3/12} = 23,4000$ psi

3.29 **e)** $\sigma_{max} = \dfrac{5,200 \times 12 \times 2}{2 \times 4^3/12} = 11,700$ psi

3.30 **e)** $V_{max} = 250. \quad \tau_{max} = \dfrac{VQ}{Ib} = \dfrac{250 \times 2 \times (1/2)}{2 \times (2^3/12) \times 2} = 93.8$

3.31 **d)** $\delta = PL^3/48EI = 500 \times 240^3 / (48 \times 30 \times 10^6 \times 2^4/12) = 3.6"$

3.32 **e)** $\delta = \dfrac{PL^3}{48EI} + \dfrac{5wL^4}{384EI} \qquad I = \dfrac{4 \times 2^3}{12} = 2.67 \text{ in}^4$

$= \dfrac{500 \times (16 \times 12)^3}{48 \times 30 \times 10^6 \times 2.67} + \dfrac{5 \times (100/12) \times (16 \times 12)^4}{384 \times 30 \times 10^6 \times 2.67} = 2.76"$

3.33 **a)** $\delta = \theta L_2 = \dfrac{PL^2}{16EI} \times L_2. \quad \therefore 4 = \dfrac{P \times 240^2 \times 120}{16 \times 30 \times 10^6 \times \pi \times 1^4/4}. \quad \therefore P = 218$ lb

3.34   **d)** $\tau_{max} = \frac{1}{2}\sqrt{\left(\sigma_x - \sigma_y\right)^2 + 4\tau^2} = \frac{1}{2}\sqrt{6,000^2 + 4 \times 4,000^2} = 5,000$ psi

3.35   **a)** $\sigma_{max} = \frac{1}{2}\left(\sigma_x + \sigma_y\right) + \tau_{max} = 0 + 4,000 = 4,000$ psi

3.36   **c)** $\tau_{max} = \frac{1}{2}\sqrt{(3,000 + 5,000)^2 + 4 \times 3,000^2} = 5,000$ psi

3.37   **d)** $\tau = Tr/J = 500 \times 12 \times 1 \Big/ \left(\pi \times 1^4/2\right) = 3,820$ psi.

$\sigma = P/A = 10,000 \Big/ \pi \times 1^2 = 3,180$ psi.

$\tau_{max} = \frac{1}{2}\sqrt{3,180^2 + 4 \times 3,820^2} = 4,137$ psi

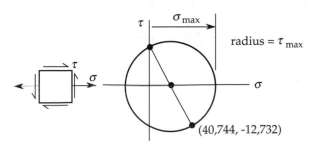

3.38   **c)** $\sigma_{max} = \frac{1}{2} \times 3,180 + 4,137 = 5,727$ psi

3.39   **b)** $M = 30,000$ in·lb. $\dfrac{My}{I} = \dfrac{30,000 \times 1}{\pi \times \left(1^4/4\right)} = 38,200$ comp.

$\dfrac{P}{A} = \dfrac{10,000}{\pi \times 1^2} = 3,183$ tens.

$\sigma_A = 38,200 - 3,183 = 35,000$ comp.

3.40   **c)** $\tau_{max} = \sigma/2 = 17,500$ psi. $VQ/Ib = 0$ on outer fibers

3.41   **c)** $\tau = \dfrac{Tr}{J} = \dfrac{2,000 \times 10 \times 1}{\pi \times \left(1^4/2\right)} = 12,732$ psi

$\sigma = \dfrac{My}{I} = \dfrac{2,000 \times 16 \times 1}{\pi \times \left(1^4/2\right)} = 40,744$ psi

$\tau_{max} = \frac{1}{2}\sqrt{40,744^2 + 4 \times 12,732^2} = 24,023$ psi

3.42   **d)** $\sigma_{max} = \dfrac{1}{2} \times 40{,}744 + 24{,}023 = 44{,}395$ psi

3.43   **c)** $M_{max} = FL_2 = 450 \times 3.28 = 1476$ ft·lb.

$$\sigma = \frac{My}{I} = \frac{(1476 \times 12) \times 1.97/2}{0.787 \times 1.97^3 / 12} = 34{,}790 \text{ psi.}$$

3.44   **d)** $\dfrac{VQ}{Ib} = \dfrac{450 \times (0.787 \times 0.985) \times 0.4925}{\left(0.787 \times 1.97^3 / 12\right) \times 0.787} = 435$ psi.

3.45   **a)**   The force F provides a torque and a force acting on the end of the steel shaft. This is best described by (a) since the top fibers experience tension. There is no normal stress in the circumferential direction, ruling out (b), (c) and (d).

3.46   **e)** $\sigma = \dfrac{My}{I} = \dfrac{(450 \times 6.56 \times 12) \times 0.985}{\pi \times 0.985^4 / 4} = 47{,}200$ psi.

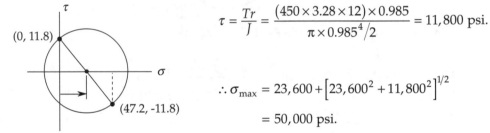

$$\tau = \frac{Tr}{J} = \frac{(450 \times 3.28 \times 12) \times 0.985}{\pi \times 0.985^4 / 2} = 11{,}800 \text{ psi.}$$

$$\therefore \sigma_{max} = 23{,}600 + \left[23{,}600^2 + 11{,}800^2\right]^{1/2}$$

$$= 50{,}000 \text{ psi.}$$

3.47   **e)** $\delta = \dfrac{PL^3}{3EI} = \dfrac{450 \times (6.56 \times 12)^3}{3 \times 30 \times 10^6 \times \pi \times 0.985^4 / 4} = 3.30$ in.

3.48   **b)** $\sigma_t = pD/2t. \quad \therefore p = 24{,}000 \times 2 \times \dfrac{1}{4} \Big/ 24 = 500$ psi

3.49   **c)** $\sigma_a = pD/4t. \quad \therefore t = \dfrac{2000 \times 48}{4 \times 30{,}000} = 0.800$ in

3.50   **e)** $\tau_{max} = \dfrac{1}{2} \sqrt{(30{,}000 - 30{,}000)^2 + 4 \times 0^2} = 0$

3.51   **a)** $\left(\dfrac{\Delta L}{L}\right)_s = \left(\dfrac{\Delta L}{L}\right)_c. \quad \therefore \varepsilon_s = \varepsilon_c. \quad \therefore \sigma_s = \dfrac{E_s}{E_c} \sigma_c = 10\,\sigma_c.$

$F_s + F_c = 400{,}000$ or $A_s \sigma_s + A_c \sigma_c = 400{,}000.$

$$\sigma_s \left[\pi\left(5.5^2 - 5^2\right) + \pi \times \frac{5^2}{10}\right] = 400{,}000. \quad \therefore \sigma_s = 16{,}430 \text{ psi}$$

3.52 d) $n = E/E_{min} = 3.$ The area is transformed :

$$I_t = \frac{4 \times 8^3}{12} - \frac{1 \times 6^3}{12} = 152.7 \text{ in}^4.$$

$$\therefore \sigma_{al} = \frac{My}{I} = \frac{2,500 \times 12 \times 4}{152.7} = 786 \text{ psi}, \quad \sigma_s = \frac{nMy}{I} = \frac{3 \times 2,500 \times 12 \times 3}{152.7} = 1770 \text{ psi}$$

3.53 d) $n = 3.$ The area is transformed :

$$I_t = \frac{12 \times 8^3}{12} - \frac{11 \times 6^3}{12} = 314 \text{ in}^4.$$

$$\therefore \sigma_s = \frac{nMy}{I} = \frac{3 \times 2,500 \times 12 \times 4}{314} = 1,146 \text{ psi}.$$

3.54 d) $60 = \frac{L}{r} = \frac{L}{\sqrt{I/A}} = \frac{L}{\sqrt{4 \times \left(4^3/12\right)/16}}.$ $\therefore L = 69.3"$ or $5.77'$

3.55 c) Assume $P = 1,000$ lb using a factor of safety of 2.

$$1,000 = \frac{\pi^2 \times 10 \times 10^6 \times \pi \times 4^4/64}{4L^2}. \quad \therefore L = 557" \text{ or } 46.4'$$

3.56 c) $P_{cr} = 4\pi^2 EI/L^2 - \alpha \Delta TEA.$ $\Delta T - \frac{4\pi^2 I}{\alpha AL^2} - \frac{4\pi^2 \times \pi \times 1^4/64}{6.5 \times 10^{-6} \times \pi \times 0.5^2 \times 120^2} = 26.4 \text{ °F}$

3.57 b) $P_{cr} = 4\pi^2 EI/L^2 = 8,000.$ $\therefore \pi^2 EI/L^2 = 2,000.$

$\therefore P_{cr} = \pi^2 EI/4L^2 = 2,000/4 = 500$ lb.

Fluid Mechanics

by Merle C. Potter

Chapter 4

Solutions to Practice Problems

4.1 **b)** a) is true for a liquid and low speed gas flows. c) and e) are true of gases. d) is true for a liquid.

Metric Units

4.2 **d)** $\tau - \mu\,du/dy$. $\therefore \mu - \tau/du/dy$. $[\mu] - \dfrac{F/L^2}{\frac{L}{T}/L} - \dfrac{FT}{L^2} - \dfrac{(ML/T^2)T}{L^2} - \dfrac{M}{LT}$

4.3 **a)** Viscosity μ varies with temperature only.

4.4 **e)** $dp = -\gamma dz = -\rho g dz$. $p = \rho RT$ (ideal gas)

$\therefore dp = -\dfrac{p}{RT}g dz$ or $\dfrac{dp}{p} = -\dfrac{g}{RT}dz$.

$\int \dfrac{dp}{p} = -\dfrac{g}{RT}\int dz$. $\therefore \ln p = -Cz$. $\therefore p = e^{-Cz}$.

4.5 **a)** $T = \tau Ar = \mu\dfrac{du}{dy}Ar = \mu\dfrac{r\omega}{t}2\pi rLr$.

$1.6 = \mu\dfrac{0.04\times1000}{0.001}2\pi\times0.04\times0.04\times0.04$. $\therefore \mu = 0.1\,\text{N}\cdot\text{s/m}^2$

4.6 **c)** $K = -V\dfrac{\Delta p}{\Delta V} = -2\dfrac{500}{-0.004} = 250\,000$ or 250 MPa.

4.7 **c)** $\sigma\, 2\pi r = \gamma \pi r^2 L$

$0.0736 \times \pi \times 0.001 = 9800\pi \times 0.0005^2\, L$

$\therefore L = 0.03$ m   or   3 cm.

4.8 **e)**   Cavitation occurs when the pressure reaches the vapor pressure = 2.45 kPa. (See Table 4.2.)

4.9 **c)** $L = V\Delta t = \sqrt{kRT}\,\Delta t = \sqrt{1.4 \times 287 \times 288} \times 1.2 = 408$ m.

Assume $T = 15°C$. T must be in absolute. $\therefore T = 288$.

4.10 **a)** $v = \dfrac{\mu}{\rho} = \dfrac{0.0034}{1.3 \times 1000} = 2.6 \times 10^{-6}$.

4.11 **a)** $p = \gamma_1 \Delta h_1 + \gamma_2 \Delta h_2$ where $\gamma_2 = SG \times \gamma_{H_2O}$

$= 9800 \times 2 + 1.04 \times 9800 \times 4 = 60\ 400$ Pa   or   60.4 kPa.

4.12 **c)** $p = \gamma h$ where $\gamma_{Hg} = 13.6\, \gamma_{H_2O}$

$= 13.6 \times 9800 \times 0.6 = 80\ 000$ Pa.

4.13 **d)** $p = \gamma h = 1.03 \times 9800 \times 1200 = 12.1 \times 10^6$ Pa   or 12.1 MPa.

4.14 **c)** $dp = -\gamma dz = -\rho g\, dz = -p\dfrac{g}{RT}\, dz$ using $p = \rho RT$

$\therefore \dfrac{dp}{p} = -\dfrac{g}{RT}\, dz$. $\int_{100}^{p} dp/p = -\dfrac{g}{RT}\int_{0}^{2000} dz$.

$\therefore \ln\dfrac{p}{100} = -\dfrac{9.8}{287 \times 293} \times 2000$. $\therefore p = 79.2$ kPa.

4.15 **c)** $F = pA$. $p = \dfrac{1000}{\pi \times 0.05^2} + 0.86 \times 9800 \times 0.2 = 129\ 000$ Pa.

$\therefore F = 129\ 000 \times \pi \times 0.0025^2 = 2.53$ N.

4.16 **a)** $p = \gamma h = (13.6 \times 9800) \times 0.2 = 26\ 700$ Pa   or 26.7 kPa.

4.17 **e)** $p + 9800 \times 0.3 = 13.6 \times 9800 \times 0.3$. $\therefore p = 37\ 000$ Pa.

4.18 **c)** $F = pA = (20\ 000 \times 9800 \times 3)\pi \times 1^2 = 155\ 000$ N.

4.19 **b)** $F = p_c A = 9800 \times \dfrac{1.5}{2} \times 1.5^2 = 16\ 500$ N.

4.20 **d)** $7P = \dfrac{5}{3}F = \dfrac{5}{3}\gamma\, h_c A$. $\therefore P = \dfrac{5}{21} \times 9800 \times 2 \times 15 = 70\ 000$ N.

4.21 **d)**   All pressures on the curved section pass through the center. Moments about the hinge give

$P = F_v = \gamma \times \text{Volume} = 9800 \times (9\pi/4) \times 4 + 9800 \times 6 \times 3 \times 4 = 983\ 000$ N.

4.22 b) $y_p = y_c + \dfrac{I_c}{y_c A} = 6 + \dfrac{5 \times 4^3/12}{6 \times 20} = 6.22$ m.

$4P = F \times 2.22 = \gamma h_c A \times 2.22.$ $\therefore P = 9800 \times 6 \times 20 \times 2.22/4 = 653\,000$ N.

4.23 e) $W = \gamma \mathcal{V}.$ $4 \times 1500 \times 9.8 = 9800 \times 6 \times 12 \times h.$ $\therefore h = 0.0833$ m.

4.24 c) $25 = 100 - 9800 \mathcal{V}.$ $\therefore \mathcal{V} = 0.00765$ m^3.
$100 = 9800(SG) \times 0.00765.$ $\therefore SG = 1.33.$

4.25 d) $\Delta p = \Delta \gamma \times h = \left(\dfrac{1}{253} - \dfrac{1}{293}\right) \times \dfrac{100}{0.287} \times 3 \times 9.8 = 5.53$ Pa.

4.26 d) $[p] = \dfrac{M}{LT^2}$ $[Q] = \dfrac{L^3}{T}$ $[D] = L$ $[\rho] = \dfrac{M}{L^3}.$ First, eliminate M, then T, then L:

$\dfrac{M}{LT^2} \cdot \dfrac{L^3}{M} \cdot \dfrac{T^2}{L^6} \cdot L^4 = p \cdot \dfrac{1}{\rho} \cdot \dfrac{1}{Q^2} \cdot D^4 = \dfrac{pD^4}{\rho Q^2}.$

4.27 a) $[\sigma] = \dfrac{M}{T^2}$ $[\rho] = \dfrac{M}{L^3}$ $[D] = L$ $[V] = \dfrac{L}{T}.$ Combine: $\dfrac{\sigma}{\rho} \dfrac{1}{V^2} \dfrac{1}{D}$

4.28 b)   Inertial force to viscous forces.

4.29 d) $(\text{Fr})_m = (\text{Fr})_p.$ $\therefore \dfrac{V_m^2}{l_m g} = \dfrac{V_p^2}{l_p g}.$ $\therefore \dfrac{V_m^2}{V_p^2} = \dfrac{1}{20}.$

$Q_m^* = Q_p^*$ or $\dfrac{Q_m}{V_m l_m^2} = \dfrac{Q_p}{V_p l_p^2}.$ $\therefore Q_m = 4 \times \dfrac{1}{20^2} \times \dfrac{1}{\sqrt{20}} = 0.0022.$

4.30 e) $\text{Re}_m = \text{Re}_p.$ $\left(\dfrac{Vl}{\nu}\right)_m = \left(\dfrac{Vl}{\nu}\right)_p.$ $\therefore \dfrac{V_m}{V_p} = \dfrac{l_p}{l_m} = 10.$
$\therefore V_m - 10\,V_p - 10 \times 10 - 100$ m/s.

4.31 a) $\text{Fr}_m = \text{Fr}_p.$ $\therefore \left(\dfrac{V^2}{lg}\right)_m = \left(\dfrac{V^2}{lg}\right)_n.$ $\therefore \dfrac{V_p^2}{V_m^2} = \dfrac{\ell_p}{\ell_m}.$

$(F_D)_m^* = (F_D)_p^*$ or $\dfrac{(F_D)_m}{\rho_m V_m^2 l_m^2} = \dfrac{(F_D)_p}{\rho_p V_p^2 l_p^2}.$ $\therefore (F_D)_p = 10\dfrac{\rho_p}{\rho_m}\dfrac{V_p^2}{V_m^2}\dfrac{l_p^2}{l_m^2} = 10\dfrac{l_p^3}{l_m^3}.$

$\therefore (F_D)_p = 10 \times 40^3 = 640\,000$ N.

4.32 d) $\text{Fr}_m = \text{Fr}_p.$ $\left(\dfrac{V^2}{lg}\right)_m = \left(\dfrac{V^2}{lg}\right)_p.$ $\therefore \dfrac{V_p^2}{V_m^2} = 10.$

$\dot{W}_p^* = \dot{W}_m^*.$ $\dfrac{\dot{W}_m}{\rho_m V_m^3 l_m^2} = \dfrac{\dot{W}_p}{\rho_p V_p^3 l_p^2}.$ $\therefore \dot{W}_p = \dfrac{V_p^3}{V_m^3}\dfrac{l_p^2}{l_m^2}\dot{W}_m = 10^3\sqrt{10} \times 20 = 63\,250$ W.

4.33 e) $V_2 = 20\pi \times 2^2/\pi \times 5^2 = 3.2$ m/s.

4.34 d) $V_2 = 20\pi \times 1^2/100 \times \pi \times 0.1^2 = 20$ m/s.

4.35 **e)** $V_2 \times 2\pi \times 40 \times 0.2 = 20 \times \pi \times 1^2. \quad \therefore V_2 = 1.25 \text{ m/s}.$

4.36 **b)** $p = \rho V^2/2 = 1.23 \times 25^2/2 = 384 \text{ Pa}. \quad F = pA = 384 \times \pi \times 0.075^2 = 6.79 \text{ N}.$

4.37 **e)** $V_2 A_2 = V_1 A_1. \quad \therefore V_2 = V_1 \times 4^2/2^2 = 4V_1.$

$$\frac{p_1}{\rho} + \frac{V_1^2}{2} = \cancel{\frac{p_2}{\rho}}^{0} + \frac{V_2^2}{2}. \quad \frac{700\,000}{1000} + \frac{V_1^2}{2} = \frac{16V_1^2}{2}. \quad \therefore V_1 = 9.66 \text{ m/s}$$

4.38 **d)** $p + \rho\dfrac{V^2}{2} + 9800 \times 0.1 = p + 13.6 \times 9800 \times 0.1.$

$\therefore V^2 = 12.6 \times 9800 \times 0.1 \times 2/1000. \quad \therefore V = 4.97 \text{ m/s}.$

4.39 **c)** Cavitation results if $p_2 = -100$ kPa.

$$\frac{p_1}{\rho} + \cancel{\frac{V_1^2}{2}}^{0} = \frac{p_2}{\rho} + \frac{V_2^2}{2}. \quad 900\,000/1000 = -100\,000/1000 + V_2^2/2. \quad \therefore V_2 = 44.7 \text{ m/s}.$$

4.40 **a)** $F = \rho A V^2 = 1.2 \times \pi \times 0.01^2 \times 80^2 = 2.41.$

4.41 **c)** $F = \rho A V^2 = .5 \times \pi \times 0.25^2 \times 1200^2 = 141\,000 \text{ N}.$

4.42 **a)** $-F = \rho A V(-V - V). \quad \therefore F = 2\rho A V^2.$

$\therefore F = 2 \times 1000 \times 0.05 \times 0.8 \times 50^2 = 200\,000 \text{ N}.$

4.43 **c)** $p_1 A_1 - F = \rho A_1 V_1(V_2 - V_1). \quad V_2 = 4V_1. \quad \dfrac{p_1}{\rho} + \dfrac{V_1^2}{2} = \cancel{\dfrac{p_2}{\rho}}^{0} + \dfrac{V_2^2}{2} = \dfrac{16V_1^2}{2}.$

$\therefore V_1 = \sqrt{\dfrac{800\,000}{1000} \times \dfrac{2}{15}} = 10.3. \qquad V_2 = 41.2.$

$\therefore F = 800\,000 \times \pi \times 0.04^2 - 1000 \times \pi \times 0.04^2 \times 10.3 \times (41.2 - 10.3) = 2420 \text{ N}.$

4.44 **b)** The energy grade line.

4.45 **a)** $\dot{W}_P = \gamma Q \left.\dfrac{\Delta p}{\gamma}\right/ 0.85 = 0.2 \times 500/0.85 = 117.6 \text{ kW}.$

4.46 **e)** $\dot{W}_T = \gamma Q \dfrac{\Delta p}{\gamma} \times 0.85 = 0.8 \times 600 \times 0.85 = 408 \text{ kW}.$

4.47 **a)** manometer: $p_1 = p_2 + \rho V_2^2/2. \quad -\dfrac{\dot{W}_T}{\gamma Q} = \dfrac{V_2^2}{2g} + \dfrac{p_2}{\gamma} - \dfrac{V_1^2}{2g} - \dfrac{p_1}{\gamma}$ (100% efficient)

$\therefore \dot{W}_T = Q\dfrac{V_2^2}{2}\rho\eta = \left(20 \times \pi \times 0.1^2\right)\dfrac{20^2}{2} \times 1000 \times .88 = 111\,000 \text{ W}.$

4.48 **b)** Increases with the velocity squared.

4.49 **d)** Increases linearly to the wall.

4.50 **c)** Vary as the 1/7th power law.

4.51 **e)** By curve E.

4.52 **b)**   By curve B.

4.53 **a)**   Pressure varies linearly $\therefore \dfrac{\Delta p}{\Delta x} = $ Const.

4.54 **e)**   The Darcy-Weisbach equation.

4.55 **d)**   Found by using loss coefficients.

4.56 **d)**   Shear stress varies linearly with radius.

4.57 **b)** $h_f = f\dfrac{L}{D}\dfrac{V^2}{2g}.$ $\therefore f = 40\dfrac{0.1}{100}\dfrac{2\times 9.8}{6^2} = 0.0218.$

4.58 **e)** $V = Q/A = \dfrac{0.02}{\pi \times 0.03^2} = 7.07$ m/s. Re $= \dfrac{VD}{\nu} = 7.07 \times 0.06/10^{-6} = 4.2\times 10^5.$

$\dfrac{e}{D} = \dfrac{.26}{60} = .0043.$ From Fig. 11.2 $f = 0.03.$

$\dot{W}_p = \dfrac{\gamma Q}{\eta}\left(\cancelto{0}{\dfrac{V_2^2}{2g}} + \cancelto{0}{\dfrac{p_2}{\gamma}} + z_2 - \cancelto{0}{\dfrac{V_1^2}{2g}} - \cancelto{0}{\dfrac{p_1}{\gamma}} - z_1 + f\dfrac{L}{D}\dfrac{V^2}{2g} + C\dfrac{V^2}{2g}\right)$

$= \dfrac{9800\times 0.02}{0.85}\left[80 - 20 + \left(0.03\dfrac{100}{.06} + 1 + .5\right)\dfrac{7.07^2}{2\times 9.8}\right] = 44\,000$ W.

4.59 **e)** $V = Q/A = \dfrac{0.006}{\pi \times 0.03^2} = 4.77$ m/s. Re $= \dfrac{VD}{\nu} = 4.77 \times .04/10^{-6} = 1.9\times 10^5.$

$\dfrac{e}{D} = \dfrac{0.046}{40} = 0.0011.$ From Fig. 4.2 $f = 0.022.$ $0 = \dfrac{p_B - p_A}{\gamma} + f\dfrac{L}{D}\dfrac{V^2}{2g} + C\dfrac{V^2}{2g}.$

$\therefore p_B = 510\,000 - \left(.022\dfrac{50}{.04} + 6.4 + 2\times .9\right)\dfrac{4.77^2}{2\times 9.8}\times 9800 = 104\,000$ Pa.

4.60 **e)** $V = Q/A = 0.3/(.15\times .4) = 5$ m/s. $R_H = \dfrac{40\times 15}{110} = 5.45.$

Re $= \dfrac{5\times 4\times .0545}{1.6\times 10^5} = 6.8\times 10^4.$ With $\dfrac{e}{D} = 0,\ f = 0.02.$

$\Delta p = f\dfrac{L}{4R_H}\dfrac{V^2}{2g}\gamma = .02\dfrac{500}{4\times .0545}\dfrac{5^2}{2}\times 1.23 = 705$ Pa.

4.61 **d)** $A_c/A_2 = .62 + .38(.5)^3 = 0.668.$ $C_1 = (1 - .668)^2 = 0.11.$

$0.11\, V_c^2/2g = C\, V_2^2/2g.$ $\therefore C = 0.11\left(\dfrac{A_2}{A_c}\right)^2 = .11\times \dfrac{1}{.668^2} = 0.25.$

4.62 **e)** $V = Q/A = 0.02/\pi \times .03^2 = 7.07$ m/s. $e/D = .15/60 = .0025$

Re $= \dfrac{7.07\times .06}{10^{-6}} = 4.2\times 10^5.$ $\therefore f = 0.024.$

$L_e = CD/f = 0.9\times 0.06/0.025 = 2.16$ m.

4.63 **a)** $Q = \dfrac{1}{n}AR_H^{2/3}S^{1/2} = \dfrac{1}{.012}\times 6\times .86^{2/3}\times .001^{1/2} = 14.3$

where $R_H = 6/(3 + 4) = 0.86$ m.

4.64 b) $Q = \dfrac{1}{n}AR_H^{2/3}S^{1/2} = \dfrac{1}{.016}4h\left(\dfrac{4h}{4+2h}\right)^{2/3}\times .001^{1/2} = 10.$

Trial - and - error : $h = 1.4$ m.

4.65 b) $Q = \dfrac{1}{n}AR_H^{2/3}S^{1/2} = \dfrac{1}{.016}\pi\times 1^2 \times .5^{2/3}S^{1/2} = 10,$ where $R_H = \dfrac{A}{P} = \dfrac{\pi\times 1^2}{2\pi} = .5.$

$\therefore S = 0.00654.$

4.66 e) $\rho = p/RT = 500/.287\times 313 = 5.57$ kg/m^3

$\dot{m} = \rho AV = 5.57\times \pi\times .05^2 \times 100 = 4.37$ kg/s

4.67 e) $T_e = T_o\left(p_e/p_o\right)^{k-1/k} = 293\left(\dfrac{100}{500}\right)^{.286} = 185$ K.

$V = Mc = 1\sqrt{1.4\times 287\times 185} = 273$ m/s .

4.68 a) $T_e = T_o\left(p_e/p_o\right)^{k-1/k} = 293\left(\dfrac{100}{800}\right)^{.286} = 162$ K . $\therefore T_e = 162 - 273 = -111°$C .

4.69 a) $M_1 = \dfrac{V_1}{c_1} = \dfrac{700}{\sqrt{1.4\times 287\times 303}} = 2.01$ $\therefore \dfrac{p_2}{p_1} = 4.54$ (from Normal Shock Table)

$\therefore p_1 = p_2/4.54 = 500/4.54 = 110$ kPa

4.70 e) $\dfrac{A}{A^*} = \dfrac{10^2}{6^2} = 2.78.$ $\therefore 2.5 < M_1 < 2.6$ (Isentropic flow Table)

4.71 b) $\sin\phi = \dfrac{1}{M} = \dfrac{1}{2}.$ $\therefore \phi = 30°$

$\tan 30° = \dfrac{1000}{L}.$ $\therefore L = 1732$ m

$\Delta t = \dfrac{L}{V} = \dfrac{1732}{2\sqrt{1.4\times 287\times 293}} = 2.52$ sec.

4.72 d) Assume $M_e = 0.3,$ the maximum if the density is assumed constant

(i.e., $\rho_e = 0.97\rho_o$). $V_e = M_ec_e = 0.3\sqrt{1.4\times 287\,T_e}.$ $\therefore V_e^2 = 36.2T_e.$

energy : $0 = \dfrac{V_e^2 - V_o^2}{2} + C_p\left(T_e - T_o\right).$

$36.2T_e = 2\times 1000\left(293 - T_e\right).$ $\therefore T_e = 287.8$

$p_o = p_e\left(T_o/T_e\right)^{\frac{k}{k-1}} = 100\left(\dfrac{293}{287.8}\right)^{\frac{1.4}{.4}} = 106.$

Solutions to Practice Problems

4.1　**b)**　a) is true for a liquid and low speed gas flows.　c) and e) are true of gases.　d) is true for a liquid.

4.2　**d)**　$\tau = \mu\, du/dy$.　$\therefore \mu = \tau/du/dy$.　$[\mu] = \dfrac{F/L^2}{\frac{L}{T}/L} = \dfrac{FT}{L^2} = \dfrac{(ML/T^2)T}{L^2} = \dfrac{M}{LT}$

4.3　**a)**　Viscosity μ varies with temperature only.

4.4　**e)**　$dp = -\gamma dz = -\rho g dz$.　$p = \rho RT$ (ideal gas)

$\therefore\ dp = -\dfrac{p}{RT} g dz$　or　$\dfrac{dp}{p} = -\dfrac{g}{RT} dz$.

$\int \dfrac{dp}{p} = -\dfrac{g}{RT} \int dz$.　$\therefore\ \ln p = -Cz$.　$\therefore\ p = e^{-Cz}$

4.5　**a)**　$T = \tau A r = \mu \dfrac{du}{dy} A r = \mu \dfrac{r\omega}{t} 2\pi r L r$.

$1.2 = \dfrac{2/12 \times 1000}{.04/12} \times 2\pi \times \dfrac{2}{12} \times \dfrac{2}{12} \times \dfrac{2}{12}\, \mu$.　$\therefore\ \mu = 8.25 \times 10^{-4}$

4.6　**c)**　$K = -\cancel{V}\, \dfrac{\Delta p}{\Delta \cancel{V}} = -60\dfrac{80}{0.12} = 40{,}000$ psi

4.7　**c)**　$\sigma 2\pi r = \gamma \pi r^2 L$

$0.005 \times 2\pi \times \dfrac{.04}{12} - 62.4\pi \dfrac{.04^2}{144} L$

$\therefore\ L - 0.0481'$ or $0.577''$

4.8　**e)**　Cavitation occurs when the pressure reaches the vapor pressure = 0.34 psi abs.　(See Table 4.1.)

4.9　**c)**　$L = V\Delta t = \sqrt{kRT}\,\Delta t = \sqrt{1.4 \times 53.3 \times 32.2 \times 530} \times 1.2 = 1354'$

Assume $T = 70°F = 530°R$

4.10　**a)**　$v = \dfrac{\mu}{\rho} = \dfrac{7.2 \times 10^{-5}}{1.3 \times 1.94} = 2.85 \times 10^{-5}$.　ρ must be in slug/ft^3.

4.11　**a)**　$p = \gamma_1 \Delta h_1 + \gamma_2 \Delta h_2 = 62.4 \times 6 + (62.4 \times 1.04) \times 12 = 1153$ psf or 8.01 psi.

4.12　**c)**　$p = \gamma h = (13.6 \times 62.4) \times 28/12 = 1980$ psf or 13.75 psi.

4.13　**d)**　$p = \gamma h = (1.03 \times 62.4) \times 4000 = 257{,}000$ psf or 1785 psi.

4.14 c) $dp = -\gamma dz = -\rho g dz = -p\dfrac{g}{RT}dz$ using $p = \rho RT$

$$\therefore \frac{dp}{p} = -\frac{g}{RT}dz. \quad \int_{14.7}^{p}dp/p = -\frac{g}{RT}\int_{0}^{6000}dz.$$

$$\therefore \ln\frac{p}{14.7} = -\frac{32.2}{53.3\times 32.2\times 530}\times 6000. \quad \therefore\ p = 11.89\text{ psia}.$$

4.15 c) $F = pA.\quad p = \dfrac{200}{\pi\times 2^2} + 0.86\times 62.4\times \dfrac{10}{12}\Big/144 = 16.2\text{ psi}.$

$$\therefore\ F = 16.2\times\pi\times\left(1/2\right)^2 = 12.74\text{ lb}.$$

4.16 a) $p = \gamma h = (13.6\times 62.4)\times\dfrac{10}{12}\Big/144 = 4.91\text{ psi}.$

4.17 e) $p + 62.4\times 1 = 13.6\times 62.4\times 1. \quad \therefore p = 786\text{ psf}$

$$\text{or}\ \ p = 5.46\text{ psi}.$$

4.18 c) $F = pA = (3\times 144 + 62.4\times 10)\pi\times 1^2 = 3318\text{ lb}.$

4.19 b) $F = \bar{p}A = 62.4\times\dfrac{5}{2}\times 5^2 = 3900\text{ lb}.$

4.20 d) $20P = \dfrac{15}{3}(62.4\times 6\times 150). \quad \therefore P = 14{,}040\text{ lb}.$

4.21 e) All pressures on the curved section pass through the center. Moments about the hinge give
$$P = F_v = \gamma\,\text{Volume} = 62.4\times 12\times\pi\,6^2\big/4 + 62.4\times 6\times 8\times 12 = 57{,}000\text{ lb}.$$

4.22 b) $y_p = \bar{y} + \dfrac{\bar{I}}{\bar{y}A} = 15 + \dfrac{15\times 10^3\big/12}{150\times 15} = 15.56'$

$$10P = 62.4\times 15\times 150\times 5.56. \quad \therefore P = 78{,}000\text{ lb}.$$

4.23 e) $W = \gamma V.\quad 4\times 3200 = 62.4\times 20\times 40h \quad\therefore h = .256'\text{ or } 3.08''.$

4.24 c) $6 = 25 - 62.4 V. \quad \therefore V = 0.3045\text{ ft}^3. \quad 25 = 62.4 S\times 0.3045. \quad \therefore S = 1.316.$

4.25 d) $\Delta p = \Delta\gamma\times h = \left(\dfrac{1}{450} - \dfrac{1}{530}\right)\dfrac{14.7\times 144}{53.3}\times 10 = 0.1332\text{ psf}.$

4.26 d) $[p] = \dfrac{M}{LT^2}\quad [Q] = \dfrac{L^3}{T}\quad [D] = L\quad [\rho] = \dfrac{M}{L^3}.$

First, eliminate M, then T, then L:

$$\frac{M}{LT^2}\cdot\frac{L^3}{M}\cdot\frac{T^2}{L^6}\cdot L^4 = p\cdot\frac{1}{\rho}\cdot\frac{1}{Q^2}\cdot D^4 = \frac{pD^4}{\rho Q^2}.$$

4.27 a) $[\sigma] = \dfrac{M}{T^2}\quad [\rho] = \dfrac{M}{L^3}\quad [D] = L\quad [V] = \dfrac{L}{T}.$

$$\frac{M}{T^2}\cdot\frac{T^2}{L^2}\cdot\frac{L^3}{M}\cdot\frac{1}{L} = \sigma\cdot\frac{1}{V^2}\cdot\frac{1}{\rho}\cdot\frac{1}{D} = \frac{\sigma}{\rho DV^2}.$$

4.28 b) Inertial force to viscous forces.

4.29 d) $(Fr)_m = (Fr)_p$. $\therefore \dfrac{V_m^2}{l_m g} = \dfrac{V_p^2}{l_p g}$. $\therefore \dfrac{V_m^2}{V_p^2} = \dfrac{1}{20}$.

$Q_m^* = Q_p^*$ or $\dfrac{Q_m}{V_m l_m^2} = \dfrac{Q_p}{V_p l_p^2}$. $\therefore Q_m = 120 \times \dfrac{1}{20^2} \times \dfrac{1}{\sqrt{20}} = .0671$ cfs.

4.30 a) $\text{Re}_m = \text{Re}_p$. $\left(\dfrac{Vl}{v}\right)_m = \left(\dfrac{Vl}{v}\right)_p$. $\therefore \dfrac{V_m}{V_p} = \dfrac{l_p}{l_m} = 10$.

$\therefore V_m = 10 V_p = 10 \times 30 = 300$ fps.

4.31 a) $Fr_m = Fr_p$. $\therefore \left(\dfrac{V^2}{\ell g}\right)_m = \left(\dfrac{V^2}{\ell g}\right)_p$. $\therefore \dfrac{V_p^2}{V_m^2} = \dfrac{\ell_p}{\ell_m}$.

$(F_D)_m^* = (F_D)_p^*$ or $\dfrac{(F_D)_m}{\rho_m V_m^2 \ell_m^2} = \dfrac{(F_D)_p}{\rho_p V_p^2 \ell_p^2}$. $\therefore (F_D)_p = 2 \dfrac{\rho_p}{\rho_m}^{1} \dfrac{V_p^2}{V_m^2} \dfrac{\ell_p^2}{\ell_m^2} = 2 \dfrac{\ell_p^3}{\ell_m^3}$.

$\therefore (F_D)_p = 2 \times 40^3 = 128{,}000$ lb.

4.32 d) $Fr_m = Fr_p$. $\left(\dfrac{V^2}{lg}\right)_m = \left(\dfrac{V^2}{lg}\right)_p$. $\therefore \dfrac{V_p^2}{V_m^2} = 10$.

$\dot{W}_p^* = \dot{W}_m^*$. or $\dfrac{\dot{W}_m}{\rho_m V_m^3 l_m^2} = \dfrac{\dot{W}_p}{\rho_p V_p^3 l_p^2}$.

$\therefore \dot{W}_p = .06 \dfrac{V_p^3}{V_m^3} \dfrac{l_p^2}{l_m^2} \dot{W}_m = .06 \times 10\sqrt{10} \times 10^2 = 189.7$ Hp.

4.33 e) $V_2 = 60\pi \times \left(\dfrac{1}{2}\right)^2 \Big/ \pi \times 1.25^2 = 9.6$ fps.

4.34 d) $V_2 = 60\pi \times \left(\dfrac{1}{2}\right)^2 \Big/ 100\pi \times .05^2 = 60$ fps.

4.35 e) $V_2 \times 0.1 \times 2\pi \times 20 = 60 \times \pi \times \left(\dfrac{1}{2}\right)^2$. $\therefore V_2 = 3.75$ fps.

4.36 b) $p = \rho V^2 / 2 = .0023 \times 90^2 / 2 = 9.32$ psf.

$F = pA = 9.32\pi \times 3^2 / 144 = 1.83$ lb.

4.37 e) $V_2 A_2 = V_1 A_1$. $\therefore V_2 = V_1 \times 2^2 / 1^2 = 4V_1$.

$\dfrac{p_1}{\rho_1} + \dfrac{V_1^2}{2} = \dfrac{p_2^{\,0}}{\rho} + \dfrac{V_2^2}{2}$. $\dfrac{100 \times 144}{1.94} + \dfrac{V_1^2}{2} = \dfrac{16 V_1^2}{2}$. $\therefore V_1 = 31.5$ fps.

4.38 d) $p + \rho \dfrac{V^2}{2} + 62.4 \times \dfrac{4}{12} = p + 13.6 \times 62.4 \times \dfrac{4}{12}$.

Using $\rho = 1.94$ slug/ft^3, $V = 16.44$ fps.

4.39 c)   Cavitation results if $p_2 = -14.7$ psi.

$$p_1/\rho + \cancel{V_1^2}^{\,0}/2 = p_2/\rho + V_2^2/2 .$$

$$150 \times 144/1.94 = -14.7 \times 144/1.94 + V_2^2/2 . \quad \therefore V_2 = 156 \text{ fps}.$$

4.40 a) $F = \rho A V^2 = .0024 \times \pi \times \dfrac{.5^2}{144} \times 200^2 = 0.524$ lb.

4.41 c) $F = \rho A V^2 = .001 \times \pi \times \dfrac{10^2}{144} \times 4000^2 = 34,900$ lb.

4.42 a) $-F = \rho A V (-V - V). \quad \therefore F = 2\rho A V^2 .$

$$\therefore F = 2 \times 1.94 \times \dfrac{2 \times 30}{144} \times 150^2 = 36,400 \text{ lb}.$$

4.43 c) $V_2 = 4V_1. \quad \dfrac{p_1}{\rho} + \dfrac{V_1^2}{2} = \cancel{\dfrac{p_2}{\rho}}^{\,0} + \dfrac{V_2^2}{2} = \dfrac{16 V_1^2}{2}. \quad \therefore V_1^2 = \dfrac{2 p_1}{15 \rho}.$

$$\therefore V_1 = \sqrt{\dfrac{2 \times 200 \times 144}{15 \times 1.94}} = 44.5. \quad \therefore V_2 = 178.$$

$$p_1 A_1 - F = \rho A_1 V_1 \left(V_2 - V_1 \right).$$

$$\therefore F = 200\pi \times 2^2 - 1.94 \times \pi \times \dfrac{2^2}{144} \times 44.5 \times 133.5 = 1508 \text{ lb}.$$

4.44 b)   The energy grade line.

4.45 a) $\dot{W}_P = \gamma Q \left. \dfrac{\Delta p}{\gamma} \right/ 0.85 = 6 \times (75 \times 144)/.85 = 76,200$ ft-lb/sec   or 139 Hp.

4.46 e) $\dot{W}_T = \gamma Q \dfrac{\Delta p}{\gamma} \times 0.85 = 3 \times (90 \times 144) \times 0.85 = 33,050$ ft-lb/sec   or 60.1 Hp.

4.47 e)   manometer : $p_1 = p_2 + \rho V_2^2 / 2$.

$$-\dfrac{\dot{W}_T}{\gamma Q} = \dfrac{\cancel{V_2^2}}{\cancel{2g}} + \dfrac{p_2}{\gamma} - \dfrac{\cancel{V_1^2}}{\cancel{2g}} - \dfrac{p_1}{\gamma} \quad (100\% \text{ efficient})$$

$$\therefore \dot{W}_T = Q \dfrac{V_2^2}{2} \rho \eta = \left(60 \times \pi \times \dfrac{4^2}{144} \right) \dfrac{60^2}{2} \times 1.94 \times .88 = 64,400 \text{ ft-lb/sec} \quad \text{or} \quad 117 \text{ Hp}.$$

4.48 b)   Increases with the velocity squared.

4.49 d)   Increases linearly to the wall.

4.50 c)   Vary as the 1/7th power law.

4.51 e)   By curve E.

4.52 b)   By curve B.

4.53 a)   Pressure varies linearly $\therefore \dfrac{\Delta p}{\Delta x} = $ Const.

4.54 e)   The Darcy-Weisbach equation.

4.55 d)   Found by using loss coefficients.

4.56 d)   Shear stress varies linearly with radius.

4.57 b) $h_L = f\dfrac{L}{d}\dfrac{V^2}{2g}. \quad \therefore f = 120\dfrac{4/12}{300}\dfrac{2\times 32.2}{20^2} = .0215$

4.58 c) $V = Q/A = \dfrac{0.6}{\pi \times 2^2/144} = 6.875$ fps.

$\text{Re} = \dfrac{VD}{\nu} = 6.875\times\dfrac{4}{10}\Big/10^{-5} = 2.3\times 10^5 \quad \dfrac{e}{D} = \dfrac{.00085}{4/12} = .0025$

from Fig. 11.2 $f = .02$

$\dot{W}_p = \dfrac{\gamma Q}{\eta}\left(\cancel{\dfrac{V_2^2}{2g}}^0 + \cancel{\dfrac{p_2}{\gamma}}^0 + z_2 - \cancel{\dfrac{V_1^2}{2g}}^0 - \cancel{\dfrac{p_1}{\gamma}}^0 - z_1 + f\dfrac{L}{D}\dfrac{V^2}{2g} + C\dfrac{V^2}{2g}\right)$

$= \dfrac{62.4\times.6}{.85}\left[200 - 50 + \left(.02\dfrac{300}{4\sqrt{12}} + 1 + .5\right)\dfrac{6.875^2}{64.4}\right] = 7230$ ft - lb sec or 13 Hp.

4.59 e) $V = Q/A = \dfrac{.02}{\pi\times 1^2/144} = 9.167$ fps

$\text{Re} = \dfrac{Vd}{\nu} = 9.167\times\dfrac{2}{12}\Big/10^{-5} = 1.53\times 10^5. \quad \dfrac{e}{d} = \dfrac{.00015}{2/12} = .0009$

From Fig. 4.2 $f = .021. \quad 0 = \dfrac{p_B - p_A}{\gamma} + f\dfrac{L}{d}\dfrac{V^2}{2g} + K\dfrac{V^2}{2g}.$

$\therefore p_B = 70 - \left(.021\dfrac{150}{2/12} + 6.4 + 2\times.9\right)\dfrac{9.167^2}{2\times 32.2}\times 62.4/144 = 54.7$ psi.

4.60 b) $V = Q/A = 40\Big/\left(\dfrac{40}{12}\times\dfrac{15}{12}\right) = 9.6$ fps. $\quad R = \dfrac{40\times 15}{110} = 5.455''.$

$\text{Re} = \dfrac{4\times 9.6\times 5.45/12}{1.6\times 10^{-4}} = 1.10\times 10^5. \quad \text{With } \dfrac{e}{d} = 0, \; f = 0.0175.$

$\gamma = p/RT = \dfrac{14.7\times 144}{53.3\times 530} = .075$

$\Delta p = f\dfrac{L}{4R}\dfrac{V^2}{2g}\gamma = .0175\dfrac{1500}{4\times 5.45/12}\dfrac{9.6^2}{64.4}\times.075 = 1.55$ psf.

4.61 d) $A_c/A_2 = .62 + .38(.5)^3 = .6675. \quad K_1 = (1 - .6675)^2 = .111.$

$.111\, V_c^2/2g = K\, V_2^2/2g. \quad \therefore K = .111(A_2/A_c)^2 = .111/.6675^2 = .249.$

4.62 e) $V = Q/A = .6\Big/\left(\pi\times 2^2/144\right) = 6.875 \quad e/d = .0005\Big/\dfrac{1}{3} = .0015$

$\text{Re} = \dfrac{6.875\times 4/12}{10^{-5}} = 2.3\times 10^5. \quad \therefore f = .022.$

$L_e = Kd/f = .9\times\dfrac{4}{12}\Big/.022 = 13.6'.$

4.63 a) $Q = \dfrac{1.49}{n}AR^{2/3}S^{1/2} = \dfrac{1.49}{.012}60\times 2.73^{2/3}\times.001^{1/2} = 460$ cfs

where $R = 60/22 = 2.73'.$

4.64 d) $Q = \dfrac{1.49}{n}AR^{2/3}S^{1/2} = \dfrac{1.49}{.016}12h\left(\dfrac{12h}{12 + 2h}\right)^{2/3}\times.001^{1/2} = 300$ cfs.

Trial - and - error : $h = 4.52'.$

4.65 a) $Q = \dfrac{1.49}{n} AR^{2/3} S^{1/2} = \dfrac{1.49}{.016}\pi \times 3^2 \times 1.5^{2/3} S^{1/2} = 100$,

where $R = \dfrac{A}{P} = \dfrac{\pi r^2}{2\pi r} = \dfrac{r}{2} = 1.5'$. $\therefore S = .00084$.

4.66 e) $\rho = p/RT = 70 \times 144/53.3 \times 32.2 \times 560 = .0105$ slug$/$ft^3

$\dot{m} = \rho AV = .0105 \times \dfrac{\pi \times 2^2}{144} \times 300 = 0.275$ slug/sec.

4.67 e) $T_e = T_o\left(p_e/p_o\right)^{k-1/k} = 530\left(\dfrac{14.7}{75}\right)^{1.4-1/1.4} = 339°\,\text{R}$.

$V = \text{M}c = 1\sqrt{1.4 \times 53.3 \times 32.2 \times 540} = 903$ fps.

4.68 a) $T_e = T_o\left(p_e/p_o\right)^{k-1/k} = 530\left(\dfrac{14.7}{120}\right)^{.2857} = 291°\,\text{R}$ or $291 - 460 = -169°\,\text{F}$.

4.69 a) $\text{M}_1 = \dfrac{V_1}{c_1} = \dfrac{2000}{\sqrt{1.4 \times 53.3 \times 32.2 \times 540}} = 1.76$. $\therefore \dfrac{p_2}{p_1} = 3.44$

(from Normal Shock Table). $\therefore p_1 = p_2\ 3.44 = 70/3.44 = 20.3$ psia

4.70 e) $\dfrac{A}{A^*} = \dfrac{4^2}{3^2} = 1.78$. $\therefore 2.0 < \text{M}_1 < 2.1$ (Isentropic flow Table).

4.71 b) $\sin\phi = \dfrac{1}{\text{M}} = \dfrac{1}{2}$. $\therefore \phi = 30°$

$\tan 30° = \dfrac{3000}{L}$. $\therefore L = 5196'$

$\Delta t = \dfrac{L}{V} = 2\dfrac{5196}{\sqrt{1.4 \times 53.3 \times 32.2 \times 530}} = 2.3$ sec.

4.72 d) Assume $\text{M}_e = 0.3$, the maximum is the density if assumed constant (i.e., $\rho_e = 0.97\rho_c$

$V_e = \text{M}_e c_e = 0.3\sqrt{1.4 \times 53.3 \times 32.2 T_e}$ $\therefore V_e^2 = 216 T_e$.

energy : $0 = \dfrac{V_e^2 - \cancel{V_o^2}^{\,0}}{2} + C_p\left(T_e - T_o\right)$.

unit conversions $\dfrac{\text{BTU}}{\text{lb}} \to \dfrac{\text{ft - lb}}{\text{slug}}$

$\therefore \dfrac{216 T_e}{2} = 0.24 \times (778 \times 32.2)(530 - T_e)$. $\therefore T_e = 520.6$

$p_o = p_e\left(T_o/T_e\right)^{\frac{k}{k-1}} = 14.7\left(\dfrac{530}{520.6}\right)^{\frac{1.4}{.4}} = 15.6$ psia.

Open Channel Flow

by David C. Wiggert

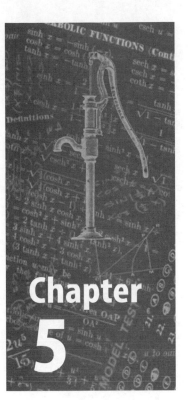

Chapter 5

Solutions to Practice Problems

5.1 Using the Chezy-Manning equation, substitute in the appropriate expressions for A and $R = A/P$:

$$Q = \frac{1.49}{n} A R^{2/3} \sqrt{S_o} = \frac{1.49}{n} \frac{A^{5/3}}{P^{2/3}} \sqrt{S_o} = \frac{1.49}{n} \frac{(by)^{5/3}}{(b+2y)^{2/3}} \sqrt{S_o}$$

Essay Problems

Substitute in known data and reduce the resulting equation:

$$500 = \frac{1.49(25y)^{5/3} \sqrt{0.0004}}{0.020(25+2y)^{2/3}}, \quad \text{or} \quad 1.568 = \frac{y^{5/3}}{(25+2y)^{2/3}}$$

Solving by trial and error, we find $y = 5.49$ ft. Now compute A and P:

$$A = 25 \times 5.49 = 137 \text{ ft}^2, \quad \text{and} \quad P = 25 + 2 \times 5.49 = 36 \text{ ft}$$

5.2 Given: $V_1 = 10$ ft/sec, $y_1 = 8$ ft, and $h = 8/12$ ft. Compute the unit discharge, energy at the upstream location 1, and the critical depth:

$$q = V_1 y_1 = 10 \times 8 = 80 \text{ ft}^2/\text{sec}, \quad E_1 = y_1 + \frac{V_1^2}{2g} = 8 + \frac{10^2}{2 \times 32.2} = 9.55 \text{ ft},$$

$$y_c = \sqrt[3]{\frac{q^2}{g}} = \sqrt[3]{\frac{80^2}{32.2}} = 5.83 \text{ ft, and} \quad E_c = \frac{3}{2} y_c = \frac{3}{2} \times 5.83 = 8.74 \text{ ft}$$

The energy equation taken from location 1 upstream of the transition to location 2 in the transition is $E_2 = E_1 - h = 9.55 - 8/12 = 8.88$ ft. Since $E_2 > E_c$, the depth at location 2 is subcritical. Therefore

$$E_2 = y_2 + \frac{q^2}{2gy_2^2}, \quad \text{or} \quad 8.88 = y_2 + \frac{80^2}{2 \times 32.2 \times y_2^2} = y_2 + \frac{99.38}{y_2^2}$$

Solving by trial and error yields $y_2 = 6.60$ ft. Hence the change in the water surface is

$$y_2 + h - y_1 = 6.6 + 0.67 - 8 = -0.73 \text{ ft}$$

5.3 At the canal entrance, the condition of critical flow is $Fr^2 = Q^2B/(gA^3) = 1$, to be solved for the unknown width b.

a) Rectangular channel: $B = b$, $A = by_c$, and

$$\therefore \frac{Q^2b}{g(by_c)^3} = \frac{Q^2}{gb^2y_c^3} = 1, \quad \text{or} \quad b = \frac{Q}{\sqrt{gy_c^3}} = \frac{635}{\sqrt{32.2 \times 3^3}} = 21.5 \text{ ft}$$

b) Trapezoidal channel:

$$B = b + 2my_c = b + 2 \times 3 \times 3 = b + 18, \quad \text{and}$$

$$A = by_c + my_c^2 = b \times 3 = 3 \times 3^2 = 3b + 27$$

$$\therefore \frac{Q^2B}{gA^3} = \frac{635^2(b+18)}{32.2 \times (3b+27)^3} = 1, \quad \text{or} \quad \frac{(b+9)^3}{(b+18)} = 464$$

Solving by trial and error gives us $b = 16.1$ ft.

5.4 a) Write the energy equation from the reservoir to section C, where critical conditions exist:

$$y_R + z_R = E_C + z_C, \quad \text{or} \quad E_C = y_R + z_R - z_C = 337 - 334 = 3 \text{ ft.}$$

Now the critical depth and specific discharge can be computed:

$$y_C = \frac{2}{3}E_C = \frac{2}{3} \times 3 = 2 \text{ ft, and} \quad q = \sqrt{gy_C^3} = \sqrt{32.2 \times 2^3} = 16.05 \text{ ft}^2/\text{sec}$$

b) The energy equation from the reservoir to section A is:

$$337 = y_A + \frac{16.05^2}{2 \times 32.2 \times y_A^2} + 330, \quad \text{or} \quad 7 = y_B + \frac{4}{y_B^2}$$

Solving by trial and error, $y_A = 6.92$ ft. In a similar fashion, the energy equation is written from the reservoir to section B, and the depth is computed to be $y_B = 4.83$ ft.

5.5 Compute the specific discharge and the downstream Froude number:

$$q = \frac{Q}{b} = \frac{400}{20} = 20 \text{ ft}^2/\text{sec}, \quad Fr_2 = \frac{q}{\sqrt{gy_2^3}} = \frac{20}{\sqrt{32.2 \times 4.5^3}} = 0.37$$

The upstream depth is

$$y_1 = \frac{y_2}{2}\left(\sqrt{1+8Fr_2^2} - 1\right) = \frac{4.5}{2}\left(\sqrt{1+8 \times 0.37^2} - 1\right) = 1.00 \text{ ft}$$

Compute the head loss across the jump:

$$h_j = y_1 + \frac{q^2}{2gy_1^2} - y_2 - \frac{q^2}{2gy_2^2} = 1.00 + \frac{400}{2 \times 32.2 \times 1.00^2} - 4.5 - \frac{400}{2 \times 32.2 \times 4.5^2} = 2.40 \text{ ft}$$

Hence the rate of energy dissipation is

$$\dot{W} = \gamma Q h_j = 62.4 \times 400 \times 2.4 = 60,000 \text{ ft-lb/sec, or } 109 \text{ horsepower}$$

5.6   Compute the specific discharges and critical depths at location 1, upstream of the transition, and at location 2 in the transition:

$$q_1 = \frac{300}{10} = 30 \text{ ft}^2/\text{sec}, \quad (y_c)_1 = \sqrt[3]{\frac{30^2}{32.2}} = 3.03 \text{ ft},$$

$$q_2 = \frac{300}{6} = 50 \text{ ft}^2/\text{sec}, \quad (y_c)_2 = \sqrt[3]{\frac{50^2}{32.2}} = 4.26 \text{ ft}$$

The critical energy at location 2 is $(E_c)_2 = 1.5(y_c)_2 = 1.5 \times 4.26 = 6.39$ ft. The specific energy at normal flow conditions is

$$E_o = y_o + \frac{q_1^2}{2gy_o^2} = 5 + \frac{30^2}{2 \times 32.2 \times 5^2} = 5.56 \text{ ft}$$

Since $E_o < (E_c)_2$, choking (a critical condition) occurs at location 2, and the depth at location 1 is greater than y_o. To compute y_1, set $E_1 = (E_c)_2$:

$$y_1 + \frac{q_1^2}{2gy_1^2} = y_1 + \frac{30^2}{2 \times 32.2 y_1^2} = y_1 + \frac{14}{y_1^2} = 6.39$$

A trial and error solution yields $y_1 = 6.0$ ft. Since $y_o > (y_c)_1$, a mild slope condition exists, and an M1 curve occurs upstream of the constriction.

5.7   First we need to compute the normal depth:

$$\frac{Qn}{1.49\sqrt{S_o}} = \frac{\left[d^2/4(\alpha - \sin\alpha \, \cos\alpha) \right]^{5/3}}{(\alpha d)^{2/3}}$$

Substituting in the given data and reducing we have

$$1.10 - \frac{(\alpha - \sin\alpha \, \cos\alpha)^{5/3}}{(\alpha)^{2/3}}$$

Solving by trial and error one finds $\alpha = 1.385$ radians, or 79 degrees. Hence the normal depth is

$$y_o = \frac{d}{2}(1 - \cos\alpha) = \frac{8}{2} \times (1 - \cos 1.385) = 3.26 \text{ ft}$$

Since $y_o > y_c$, a mild slope condition exists. The water surface consists of an M3 curve beginning at the inlet, followed by a hydraulic jump to an M2 curve, which terminates at critical depth at the outlet.

5.8   Compute the specific discharge and critical depth:

$$q = 300/15 = 20 \text{ ft}^2/\text{sec}, \quad \text{and} \quad y_c = \sqrt[3]{20^2/32.2} = 2.31 \text{ ft}$$

Compute the normal depth using the Chezy-Manning equation:

$$\frac{Qn}{1.49\sqrt{S_o}} = \frac{300 \times 0.014}{1.49 \times \sqrt{0.0008}} = 99.7 = \frac{(15y_o)^{5/3}}{(15 + 2y_o)^{2/3}}$$

Solving by trial and error, $y_o = 3.65$ ft. Since $y_o > y_c$, mild channel conditions exist. Upstream of the gate there will be an M1 profile, with an M3 profile downstream of the gate terminating in a hydraulic jump to normal flow conditions.

5.9 Evaluate Q using the varied flow equation. First evaluate the specific energies and the slope of the energy grade line:

$$E_1 = y_1 + \frac{Q^2}{2gA_1^2} = 3.5 + \frac{Q^2}{2 \times 32.2 \times (8 \times 3.5)^2} = 3.5 + \frac{Q^2}{50490} \ ,$$

$$E_2 = y_2 + \frac{Q^2}{2gA_2^2} = 4 + \frac{Q^2}{2 \times 32.2 \times (8 \times 4)^2} = 4 + \frac{Q^2}{65950} \ ,$$

$$y_m = \frac{1}{2}(y_1 + y_2) = \frac{1}{2}(3.5 + 4) = 3.75 \text{ ft},$$

$$S = \left(\frac{Qn}{1.49}\right)^2 \frac{(b + 2y_m)^{4/3}}{(by_m)^{10/3}} = \frac{Q^2 \times 0.013^2 \times (8 + 2 \times 3.75)^{4/3}}{1.49^2 (8 \times 3.75)^{10/3}} = 3.507 \times 10^{-8} Q^2$$

Substitute into the varied-flow equation $\Delta x(S_o - S) = E_2 - E_1$:

$$200\left(0.005 - 3.507 \times 10^{-8} Q^2\right) = 4 + \frac{Q^2}{65950} - 3.5 - \frac{Q^2}{50490} \ , \quad \text{or}$$

$$2.37 \times 10^{-6} Q^2 = 0.5$$

Hence $Q = 459 \text{ ft}^3/\text{sec}$.

5.10 a) Compute specific discharge and critical depth:

$$q = 1200 / 40 = 30 \text{ ft}^2/\text{sec} , \quad \text{and} \quad y_c = \sqrt[3]{30^2 / 32.2} = 3.03 \text{ ft}$$

Since $y_o > y_c$, mild flow conditions prevail. From Table 5.4, an M1 profile exists upstream of the dam.

b) Use the step method to evaluate the M1 profile. First set up the functions S and E:

$$S = \left(\frac{Qn}{1.49}\right)^2 \frac{1}{A^2 R^{4/3}} = \frac{145.9}{A^2 R^{4/3}} , \quad A = 40y , \quad R = \frac{40y}{40 + 2y}$$

$$E = y + \frac{q^2}{2gy^2} = y + \frac{30^2}{2 \times 32.2 \times y^2} = y + \frac{14}{y^2}$$

Then use the relation $x_{i+1} = x_i + (E_{i+1} - E_i)/(S_o - S) = x_i + (E_{i+1} - E_i)/(0.0008 - S)$ to complete the M1 profile evaluation. The results are tabulated below:

i	y_i	E_i	y_m	$A(y_m)$	$R(y_m)$	$S(y_m)$	x_i
1	8.50	8.69					0
2	7.50	7.75	8.00	320	5.71	0.000140	-1431
3	6.50	6.83	7.00	280	5.19	0.000207	-2979
4	5.50	5.96	6.00	240	4.62	0.000330	-4826
5	4.50	5.19	5.00	200	4.00	0.000575	-8251

Note that since we are computing the M1 profile upstream from the starting depth, the x-values are negative. The approximate location where normal depth is reached is 8250 ft upstream of the dam.

5.11 **c)** The critical depth is $y_c = \sqrt[3]{(680/60)^2 / 32.2} = 1.58$ ft.

5.12 **a)** Set up the Chezy-Manning equation:

$$680 = \frac{1.49}{0.016}(60y_o)\left(\frac{60y_o}{60+2y_o}\right)^{2/3}\sqrt{0.0005} = 1917 \times \frac{(y_o)^{5/3}}{(60+2y_o)^{2/3}}$$

Solving by trial and error, $y_o = 2.87$ ft.

5.13 **d)** Use the hydraulic jump equation to evaluate y_1. The downstream Froude number is

$$V_2 = \frac{680}{2.5 \times 60} = 4.53 \text{ ft/sec}, \quad Fr_2 = \frac{4.53}{\sqrt{32.2 \times 2.5}} = 0.505$$

$$\therefore y_1 = \frac{2.5}{2}\left(\sqrt{1 + 8 \times 0.505^2} - 1\right) = 0.93 \text{ ft}.$$

5.14 **e)** First find the energy loss across the jump:

$$q = \frac{680}{60} = 11.33 \text{ ft}^2/\text{sec}$$

$$h_j = E_1 - E_2 = 0.93 + \frac{11.33^2}{2 \times 32.2 \times 0.93^2} - 2.5 - \frac{11.33^2}{2 \times 32.2 \times 2.5^2} = 0.42 \text{ ft}$$

The power dissipated is $\dot{W} = \gamma Q h_j / 550 = 62.4 \times 680 \times 0.42 / 550 = 32.4$ Hp.

5.15 **d)** Use Eq. 5.3.2 to estimate the distance on the apron from the toe of the dam to immediately upstream of the jump (location 1):

$$E_{toe} = 0.35 + \frac{11.33^2}{2 \times 32.2 \times 0.35^2} = 16.62 \text{ ft}, \quad E_1 = 0.9 + \frac{11.33^2}{2 \times 32.2 \times 0.9^2} = 3.36 \text{ ft},$$

$$y_m = \frac{1}{2}(0.35 + 0.9) = 0.625 \text{ ft}, \quad A = 60 \times 0.625 = 37.5 \text{ ft}^2,$$

$$P = 60 + 2 \times .625 = 61.2 \text{ ft}, \quad S = \left(\frac{680 \times 0.014}{1.49 \times 37.5}\right)^2 \frac{1}{(37.5/61.2)^{4/3}} = 0.0558,$$

$$\therefore \text{ Profile length} = \frac{3.36 - 16.2}{0.0005 - 0.0558} = 232 \text{ ft}$$

The length of the jump is $6 \times 2.5 = 15$ ft. Hence the required length of the apron is $L = 232 + 15 = 247$ ft, or about 250 ft. (Normal design practice would require additional length as a safety factor.)

5.16 **b)** In a manner similar to Problem 12, the normal depth on the apron is $y_o = 2.65$ ft. Since y_o is greater than y_c in both the river and apron ($y_o = 2.87$ ft in river from Prob. 12), mild slope conditions prevail. Ahead of the dam, the depth is $h + y_c = 20 + 1.6 = 21.6$ ft, hence the profile there is an

M1 curve. On the apron, between the toe and the hydraulic jump, an M3 curve exists.

5.17 **e)** Use the Chezy-Manning equation to determine the discharge:

$$A = 10 \times 3 + 2 \times 3^2 = 48 \text{ ft}^2, \quad P = 10 + 2 \times 3 \times \sqrt{1+2^2} = 23.4 \text{ ft}, \quad R = \frac{48}{23.4} = 2.05 \text{ ft}$$

$$\therefore \quad Q = \frac{1.49}{0.012} \times 48 \times 2.05^{2/3} \times \sqrt{0.0002} = 136 \text{ ft}^3/\text{sec}$$

5.18 **a)** The conduits are flowing half full. For each conduit the area and hydraulic radius are $A = \pi d^2/8$, and $R = d/4$. Apply the Chezy-Manning equation to one of the conduits using one-half of the discharge:

$$68 = \frac{1.49}{0.013} \times \frac{\pi d^2}{8} \times \left(\frac{d}{4}\right)^{2/3} \times \sqrt{0.0002} = 0.252 d^{8/3}, \quad \therefore \quad d = 8.2 \text{ ft}$$

The diameter to the nearest foot is $d = 8$ ft.

5.19 **c)** Use the Chezy-Manning equation to evaluate the normal depth:

$$136 = \frac{1.49}{0.012} \frac{A^{5/3}}{P^{2/3}} \sqrt{0.0003}, \quad \text{which reduces to} \quad 63.2 = \frac{\left(10 y_o + 2 y_o^2\right)^{5/3}}{\left(10 + 2\sqrt{5} y_o\right)^{2/3}}$$

A trial and error solution yields $y_o = 2.7$ ft.

5.20 **d)** Use Eq. 5.2.3 (assume that there are rectangular side walls in the vicinity of the weir):

$$136 = 3.087 \times 25 \times Y^{3/2}, \quad \therefore \quad Y = 1.46 \text{ ft}.$$

Flow in Piping Systems

by David C. Wiggert

Solutions to Problems

6.1 This is a category 1 problem.

Essay Problems

a) The kinematic viscosity of 60° F water is 1.22×10^{-5} ft^2/sec. From Table 6.1 we find

$e = 0.00085$ ft. Compute the velocity, Reynolds number, and relative roughness:

$$D = 1.5/12 = 0.125 \text{ ft}, \quad V = \frac{27/(7.48 \times 60)}{\pi \times 0.125^2/4} = 4.9 \text{ ft/sec}.$$

$$Re = \frac{4.9 \times 0.125}{1.22 \times 10^{-5}} = 5.02 \times 10^4, \quad \varepsilon = \frac{0.00085}{0.125} = 0.0068$$

Compute the friction factor:

$$f = 1.325\left[\ln\left(0.27 \times 0.0068 + 5.74 \times (5.02 \times 10^4)^{-0.9}\right)\right]^{-2} = 0.035$$

(One could also use the Moody Diagram to estimate the friction factor.) Substitute into the Darcy-Weisbach equation:

$$h_L = \frac{0.035 \times 350 \times 4.9^2}{0.125 \times 2 \times 32.2} = 36.8 \text{ ft}, \quad \text{or} \quad 36.8 \times \frac{62.4}{144} = 15.9 \text{ psi}$$

b) From Table 6.2, assume $C = 110$. Then substitute known data into the Hazen-Williams formula:

$$h_L = \frac{4.72 \times 350 \times [27/(7.48 \times 60)]^{1.85}}{110^{1.85} \times 0.125^{4.87}} = 38.1 \text{ ft}, \quad \text{or} \quad 38.1 \times \frac{62.4}{144} = 16.5 \text{ psi}$$

Note the difference in the two results; both are acceptable answers for most problems. Solution (a), however, is more accurate.

6.2   This is a category 2 problem. The head loss due to friction is $h_L = 5 \times 144 / 62.4 = 11.5$ ft. Substitute into the Darcy-Weisbach equation:

$$11.5 = f \frac{600}{4} \times \frac{V^2}{2 \times 32.2}, \quad \text{or} \quad fV^2 = 4.94$$

The Moody diagram is used to estimate f. The kinematic viscosity is $v = 1.22 \times 10^{-5}$ ft^2/sec, and the relative roughness is $\varepsilon = 0.01/4 = 0.0025$. First assume $f = 0.03$. Then $V = \sqrt{4.94/0.03} = 12.8$ ft/sec, $Re = 12.8 \times 4/1.22 \times 10^{-5} = 4.2 \times 10^6$, and from the Moody diagram, we find the friction factor to be $f = 0.025$. Hence the new estimate of the velocity is $V = 14.1$ ft/sec, and $Re = 14.1 \times 4/1.22 \times 10^{-5} = 4.6 \times 10^6$. The friction factor remains at 0.025, so that the final velocity is $V = 14.1$ ft/sec, and the discharge is $Q = 14.1 \times \pi \times 4^2/4 = 177$ ft^3/sec.

As an alternative to the trial and error approach, the empirical formula for a category 2 problem, Table 6.3, can be used since no pump is involved:

$$Q = -0.965 \times \left(\frac{32.2 \times 4^5 \times 11.5}{600} \right)^{0.5} \ln \left[\frac{0.01}{3.7 \times 4} + \left(\frac{3.17 \times (1.22 \times 10^{-5})^2 \times 600}{32.2 \times 4^3 \times 11.5} \right)^{0.5} \right]$$

$$= 177 \text{ ft}^3/\text{sec}$$

The empirical formula gives an answer that is as precise as the trial and error solution.

6.3   This is a category 3 pipe problem. The following data are given: $L = 1000$ ft, $Q = 20/(7.48 \times 60) = 0.0446$ ft^3/sec, and $h_L = 210$ ft. Substitute into the Darcy-Weisbach equation and simplify:

$$210 = \frac{8 \times f \times 1000 \times 0.0446^2}{32.2 \times \pi^2 \times D^5}, \quad \text{or} \quad D = 0.189 f^{1/5}$$

This relation will be used in conjunction with the Moody diagram to estimate the friction factor. For drawn tubing, $e = 5 \times 10^{-6}$ ft from Table 6.1, and the kinematic viscosity for water at 80° F is $v = 0.93 \times 10^{-5}$ ft^2/sec. The two parameters required for using the diagram are

$$\varepsilon = \frac{e}{D} = \frac{5 \times 10^{-6}}{D}, \quad \text{and} \quad Re = \frac{4Q}{\pi D v} = \frac{4 \times 0.0446}{\pi \times D \times 0.93 \times 10^{-5}} = \frac{6110}{D}$$

Begin the iterative solution by assuming that $f = 0.015$. Then

$$D = 0.189 \times 0.015^{1/5} = 0.0816 \text{ ft}, \quad \varepsilon = \frac{5 \times 10^{-6}}{0.0816} = 6 \times 10^{-5}, \quad Re = \frac{6110}{0.0816} = 75,000$$

The updated friction factor read from the Moody diagram is $f = 0.019$. Repeating the calculations one finds that $D = 0.0855$ ft, $\varepsilon = 6 \times 10^{-5}$, and $Re = 71,000$. From the Moody diagram, the friction factor again is $f - 0.019$, unchanged from the previous estimate. Hence $D = 0.0855 \times 12 = 1.03$ in. One could use either a 1-inch-diameter pipe or a 2-inch-diameter pipe. A 1-inch-diameter pipe will yield a head loss of $h_L = 237$ ft, and a 2-inch-diameter pipe will yield $h_L = 74$ ft.

The solution using the formula for a category 3 problem in Table 6.3 is

$$D = 0.66 \left[(5 \times 10^{-6})^{1.25} \left(\frac{1,000 \times 0.0446^2}{32.2 \times 210} \right)^{4.75} + 0.93 \times 10^{-5} \times 0.0446^{9.4} \left(\frac{1,000}{32.2 \times 210} \right)^{5.2} \right]^{0.04}$$

$$= 0.0868 \text{ ft, or } 1.04 \text{ in}$$

Note that the solution is as accurate as using the Moody diagram with the trial and error procedure.

6.4   The kinematic viscosity is $v = 1.22 \times 10^{-5}$ ft^2/sec, the relative roughness is $\varepsilon = 0.00015/0.167 = 0.0009$, and the pipe diameter is $D = 2/12 = 0.167$ ft. Assume a completely turbulent flow regime, so that the friction factor can be approximated by $f = 1.325[\ln(0.27 \times 0.0009)]^{-2} = 0.019$. Next, write the energy equation from the reservoir surface to the pipe outlet:

$$140 = \frac{V^2}{2g} + \left[\frac{fL}{D} + 2K_{elbow} + K_{entrance} \right] \frac{V^2}{2g}$$

Assume standard elbows so that $K_{elbow} = 0.9$, and with $K_{entrance} = 0.5$ from Table 4.4, substitute known data into the energy equation:

$$140 = \left[1 + \frac{0.019 \times 400}{0.167} + 2 \times 0.9 + 0.5 \right] \frac{V^2}{2 \times 32.2} \quad . \quad \therefore \ V = 13.6 \text{ ft/sec}$$

We perform one more iteration to determine a better estimate of the velocity:

$$Re = \frac{13.6 \times 0.167}{1.22 \times 10^{-5}} = 1.9 \times 10^5 \ ,$$

$$f = 1.325 \left[\ln \left(0.27 \times 0.0009 + 5.74(1.9 \times 10^5)^{-0.9} \right) \right]^{-2} = 0.021$$

Substituting the new value of f into the energy equation and solving we find that $V = 13.0$ ft/sec. Hence $Q = 13.0 \times \pi \times 0.167^2/4 = 0.285$ ft^3/sec. The hydraulic grade line has sudden drops at the entrance and elbows, and a linear drop over the length of the pipe.

6.5   The energy equation for the system is $H_P = (h_L)_{sys} + 150$, or

$$\frac{\dot{W}_P \eta}{\gamma Q_{sys}} = (R_1 + R_2 + R_3) Q_{sys}^2 + 150 \ , \text{ in which } R_i = \frac{1}{2gA_i^2} \left(\frac{fL}{D} + \Sigma K \right)_i$$

Compute the R-values:

$$A_1 = \frac{\pi}{4} \times 5^2 = 19.6 \text{ ft}^2, \quad R_1 = \frac{1}{2 \times 32.2 \times 19.6^2}\left(\frac{0.018 \times 600}{5} + 2\right) = 0.00017$$

$$A_2 = \frac{\pi}{4} \times 3^2 = 7.1 \text{ ft}^2, \quad R_2 = \frac{1}{2 \times 32.2 \times 7.1^2}\left(\frac{0.020 \times 1000}{3} + 0\right) = 0.00205$$

$$A_3 = \frac{\pi}{4} \times 4^2 = 12.6 \text{ ft}^2, \quad R_3 = \frac{1}{2 \times 32.2 \times 12.6^2}\left(\frac{0.019 \times 400}{4} + 10\right) = 0.00116$$

Substituting the known data into the energy equation we have

$$\frac{2500 \times 550 \times 0.82}{62.4 Q_{sys}} = (0.00017 + 0.00205 + 0.00116)Q_{sys}^2 + 150$$

which reduces to $18,070 = 0.00338 Q_{sys}^3 + 150 Q_{sys}$. Solving the equation by trial and error gives

$$Q_{sys} = 99 \text{ ft}^3 / \text{sec}.$$

6.6 Compute the R-values:

Pipe 1: $L_e = \frac{2 \times 0.333}{0.02} = 33$, $\quad R_1 = \frac{8 \times 0.02 \times 233}{32.2 \times \pi^2 \times 0.333^5} = 29$

Pipe 2: $L_e = 0$, $\qquad\qquad R_2 = \frac{8 \times 0.02 \times 300}{32.2 \times \pi^2 \times 0.5^5} = 5$

Pipe 3: $L_e = \frac{2 \times 0.167}{0.02} = 33$, $\quad R_3 = \frac{8 \times 0.02 \times 213}{32.2 \times \pi^2 \times 0.167^5} = 826$

Pipe 4: $L_e = \frac{1 \times 0.333}{0.02} = 17$, $\quad R_4 = \frac{8 \times 0.02 \times 277}{32.2 \times \pi^2 \times 0.333^5} = 34$

We next find the system head loss, and then the discharge in each pipe:

$$(h_L)_{sys} = \frac{5^2}{\left(1/\sqrt{29} + 1/\sqrt{5} + 1/\sqrt{826} + 1/\sqrt{34}\right)^2} = 35.5 \text{ ft}$$

$$Q_1 = \sqrt{35.5/29} = 1.11 \text{ ft}^3/\text{sec}, \quad Q_2 = \sqrt{35.5/5} = 2.66 \text{ ft}^3/\text{sec}$$

$$Q_3 = \sqrt{35.5/826} = 0.21 \text{ ft}^3/\text{sec}, \quad Q_4 = \sqrt{35.5/34} = 1.02 \text{ ft}^3/\text{sec}$$

The sum of the four discharges is equal to the system discharge.

6.7 Compute the loss terms for each pipe:

Pipe 1: $L_e = \frac{2 \times 0.5}{0.02} = 50$, $\quad (h_L)_1 = \frac{8 \times 0.02 \times 2550}{32.2 \times \pi^2 \times 0.5^5} Q_1^2 = 41 Q_1^2$

Pipe 2: $L_e = \frac{3 \times 0.333}{0.025} = 40$, $\quad (h_L)_2 = \frac{8 \times 0.025 \times 1640}{32.2 \times \pi^2 \times 0.333^5} Q_2^2 = 252 Q_2^2$

Pipe 3: $L_e = \frac{7 \times 0.5}{0.018} = 190$, $\quad (h_L)_3 = \frac{8 \times 0.018 \times 3490}{32.2 \times \pi^2 \times 0.5^5} Q_3^2 = 51 Q_3^2$

Assume the flow directions shown in Fig. 6.4a, and write the energy relations for each pipe:

Pipe 1: $H_B = 65 - 41 Q_1^2$, or $Q_1 = \sqrt{(65 - H_B)/41}$

Pipe 2: $H_B = 5 + 252 Q_2^2$, or $Q_2 = \sqrt{(H_B - 5)/252}$

Pipe 3: $H_B = 43 + 51 Q_3^2$, or $Q_3 = \sqrt{(H_B - 43)/51}$

The solution is found by iterating values of H_B, computing Q_1, Q_2, and Q_3, and checking the continuity balance at the junction: $\Sigma Q = Q_1 - Q_2 - Q_3$. The results are tabulated below; five iterations are performed until the continuity balance is satisfied within a tolerable limit.

Iteration	H_B	Q_1	Q_2	Q_3	ΣQ
1	43	0.732	0.388	0	+0.344
2	45	0.698	0.398	0.198	+0.102
3	46	0.681	0.403	0.242	+0.036
4	47	0.663	0.408	0.280	-0.025
5	46.6*	0.670	0.406	0.266	-0.002

* Estimated by linear interpolation between the two previous values of H_B and Q .

Hence the discharges are Q_1 = 0.67 ft³/sec, Q_2 = 0.40 ft³/sec, and Q_3 = 0.27 ft³/sec.

6.8   With $D = 8/12=0.668$ ft, assuming a constant friction factor, and recognizing that $z_2 - z_1 = 105$ ft, the demand curve can be constructed:

$$H_P = 105 + \left(\frac{0.02 \times 8,000}{0.667} + 12.5 \right) \frac{Q^2}{2 \times 32.2 \times \left(\pi \times 0.667^2 / 4 \right)^2}$$

$$= 105 + 32Q^2$$

a) Assume $Q = 1.0$ ft³/sec, then $H_p = 105 + 32 \times 1^2 = 137$ ft. On the pump characteristic curve of Fig. 6.5, $H_P \cong 175$ ft at $Q = 1.0$ ft³/sec. Try another discharge, say $Q = 1.5$ ft³/sec; then from the demand curve $H_P = 177$ ft, and from the pump curve $H_P \cong 170$ ft. Finally, with $Q = 1.4$ ft³/sec, from the demand curve $H_P = 168$ ft, and from the pump curve $H_P \cong 170$ ft. Thus $Q = 1.4$ ft³/sec and $H_P = 170$ ft.

b) From Fig 6.5, the pump efficiency at $Q = 1.4$ ft³/sec is approximately 0.7. Hence, the required power input to the pump is $\dot{W}_P = 53 \times 1.4 \times 170 / 0.7 = 18,000$ ft - lb/sec, or 32.8 horsepower.

6.9   First compute the head loss in the suction pipe. Neglecting minor losses (no information is given) we have, see Eq. 6.1.1 and use Fig. 4.2,

$$h_L = \frac{8 \times 0.02 \times 16 \times (13/7.48)^2}{32.2 \times \pi^2 \times 0.5^5} = 0.78 \text{ ft}$$

a) For water at 80° F, $\gamma = 62.2$ lb/ft³, and $p_{vap} = 0.507$ psi. Substituting into Eq. 6.3.2, the condition for imminent cavitation is

$$NPSH = \frac{(15 - 0.507) \times 144}{62.2} - 16 - 0.78 = 16.8 \text{ ft}$$

b) With $p_{atm} = 12$ psi and the same flow conditions for part a), use Eq. 6.3.2 to determine the required z:

$$\Delta z = \frac{p_{atm} - p_{vap}}{\gamma} - h_L - NPSH = \frac{(12 - 0.507) \times 144}{62.2} - 0.78 - 16.8 = 9.0 \text{ ft}$$

Hence the change in elevation is $16 - 9 = 7$ ft.

6.10 The following parameters are given: $N_1 = 970$ rpm, $N_2 = 1{,}200$ rpm, $H_1 = 38$ ft, $Q_1 = 230 \text{ ft}^3/\text{sec}$, and $D_2 : D_1 = 1 : 2$. Using the similitude relations, Eqs. 6.3.4 and 6.3.5, compute Q_2 and H_2 :

$$H_2 = 38 \times \left(\frac{1{,}200}{970}\right)^2 \times \left(\frac{1}{2}\right)^2 = 14.5 \text{ ft}$$

$$Q_2 = 230 \times \frac{1{,}200}{970} \times \left(\frac{1}{2}\right)^3 = 35.6 \text{ ft}^3/\text{sec}$$

The power is $\dot{W}_2 = 62.4 \times 35.6 \times 14.5 / 0.85 = 37{,}900$ ft - lb/sec, or 69 horsepower.

Multiple Choice Problems

Problems 6.11-20

Before answering the specific questions, it is expedient to compute the R-values for each pipe (recall that $h_L = RQ^2$):

Pipe 1 : $Le = \dfrac{1 \times 0.333}{0.02} = 17$ ft , $R_1 = \dfrac{8 \times 0.02 \times 27}{32.2 \times \pi^2 \times 0.333^5} = 3$

Pipe 2 : $Le = \dfrac{2 \times 0.333}{0.02} = 33$ ft , $R_2 = \dfrac{8 \times 0.02 \times 533}{32.2 \times \pi^2 \times 0.333^5} = 66$

Pipe 3 : $Le = \dfrac{2 \times 0.333}{0.02} = 33$ ft , $R_3 = \dfrac{8 \times 0.02 \times 2{,}033}{32.2 \times \pi^2 \times 0.333^5} = 250$

Pipe 4 : $Le = \dfrac{4 \times 0.333}{0.02} = 67$ ft , $R_4 = \dfrac{8 \times 0.02 \times 1{,}817}{32.2 \times \pi^2 \times 0.333^5} = 224$

6.11 **d)** Compute the discharge and head loss in pipe 2:

$$Q_2 = 5{,}000 \frac{\text{gal}}{\text{hr}} \times \frac{1 \text{ ft}^3}{7.48 \text{ gal}} \times \frac{1 \text{ hr}}{3600 \text{ sec}} = 0.186 \frac{\text{ft}^3}{\text{sec}},$$

$$(h_L)_2 = R_2 Q_2^2 = 66 \times 0.186^2 = 2.28 \text{ ft}$$

6.12 **b)** The flow path is pipe 1 - pipe 2 - pipe 3; there is no flow in pipe 4. The demand curve from A to B is

$$H_P = z_B - z_A + (h_L)_1 + (h_L)_2 + (h_L)_3 = 430 + (3 + 66 + 250)Q^2 = 430 + 319Q^2$$

A trial and error solution is employed: assume Q, compute H_P from the demand curve, and compare it with H_P from the pump characteristic curve.

Trial 1: $Q = 10{,}000$ gal/hr $= 0.371 \text{ ft}^3/\text{sec}$,
H_P (demand) $= 430 + 319 \times 0.371^2 = 474$ ft,
H_P (pump) $= 465$ ft

Trial 2: $Q = 9{,}000$ gal/hr $= 0.334$ ft^3/sec,

H_P (demand) $= 430 + 319 \times 0.334^2 = 467$ ft,

H_P (pump) $= 466$ ft

6.13 **a)** $Q_2 = 11{,}000$ gal/min $= 0.409$ ft^3/sec, and from the pump curve, $H_P = 460$ ft. Write the energy equation from A to C:

$$H_C = H_P - (R_1 + R_2)Q_2^2 = 460 - 69 \times 0.409^2 = 449 \text{ ft}$$

The energy equation from C to D is $H_C = z_D + R_4 Q_4^2$, where H_C is the hydraulic grade line at location C. Therefore the discharge in pipe 4 is

$$Q_4 = \sqrt{\frac{H_C - z_D}{R_4}} = \sqrt{\frac{449 - 445}{224}} = 0.134 \text{ ft}^3/\text{sec, or } 3{,}600 \text{ gal/hr.}$$

6.14 **d)**   The discharge is $Q = 8{,}000/(7.48 \times 3600) = 0.297$ ft^3/sec. From the pump characteristic curve,

$H_P = 470$ ft, and $\eta = 0.8$. Therefore the required power is

$$\dot{W}_P = \gamma Q H_P / \eta = 62.4 \times 0.297 \times 470 / 0.8 = 10{,}900 \text{ ft - lb/sec, or } 19.8$$
horsepower.

6.15 **c)**   First determine the head loss in the suction pipe:

$$Q = \frac{10{,}000}{7.48 \times 3600} = 0.371 \text{ ft}^3/\text{sec} , \quad (h_L)_1 = 3 \times 0.371^2 = 0.41 \text{ ft}$$

Using Eq. 6.3.2, the required pump height is:

$$\Delta z = \frac{p_{atm}}{\gamma} \frac{p_{vap}}{} - (h_L)_1 - NPSH = \frac{(14.7 - 0.256) \times 144}{62.4} - 0.41 - 25 = 7.9 \text{ ft}$$

6.16 **b)** The discharge through the pump is $Q = 12{,}000/(7.48 \times 3600) = 0.446$ ft^3/sec, and from the characteristic curve the pump head is $H_P = 454$ ft. The energy equation from A to C is $H_P = z_C + p_C/\gamma - z_A + (h_L)_1 + (h_L)_2$. Therefore

$$\frac{p_C}{\gamma} = 454 - 300 - (3 + 66) \times 0.446^2 = 140 \text{ ft} , \quad \text{and } p_C = \frac{62.4 \times 140}{144} = 61 \text{ lb/in}^2$$

6.17 **b)** Write the energy equation from C to D and solve for the pressure at location D:

$$p_C/\gamma + z_C = p_D/\gamma + z_D$$

(Since there is no flow in pipe 4, the energy equation reduces to the law of hydrostatics.) From Problem 16, the pressure at location C is 61 lb/in^2. Therefore,

$$p_D = 61 + 62.4 \times \frac{300 - 445}{144} = -1.83 \text{ lb/in}^2$$

6.18   **d)** From the pump curve, for $H_P = 460$ ft, $Q = 11,000$ gal/hr, or 0.409 ft^3/sec. Write the energy equation from A to C and evaluate the hydraulic grade line:

$$H_C = H_P - (h_L)_1 - (h_L)_2 = 460 - 69 \times 0.409^2 = 449 \text{ ft}$$

Thus the discharge in pipe 3, between location C and location B is

$$Q_3 = \sqrt{\frac{(H_C - H_B)}{R_3}} = \sqrt{\frac{449 - 430}{250}} = 0.276 \text{ ft}^3/\text{sec}, \text{ or } 7,400 \text{ gal/hr}$$

6.19   **c)** The flow path is pipe 1 - pipe 2 - pipe 4; there is no flow in pipe 3. Write the energy equation from location A to location D:

$$H_P = z_D - z_A + (h_L)_1 + (h_L)_2 + (h_L)_4 = 445 + (3 + 66 + 224)Q^2 = 445 + 293Q^2$$

A trial and error solution is employed: assume Q, compute H_P from the demand curve, and compare it with H_P from the pump characteristic curve.

Trial 1: $Q = 10,000$ gal/hr $= 0.371$ ft^3/sec,
H_P (demand) $= 445 + 293 \times 0.371^2 = 485$ ft,
$\quad\quad H_P$ (pump) $= 465$ ft

Trial 2: $Q = 7,500$ gal/hr $= 0.278$ ft^3/sec,
H_P (demand) $= 445 + 293 \times 0.278^2 = 468$ ft,
$\quad\quad H_P$ (pump) $= 475$ ft

Trial 3: $Q = 8,000$ gal/hr $= 0.297$ ft^3/sec,
H_P (demand) $= 445 + 293 \times 0.297^2 = 471$ ft,
$\quad\quad H_P$ (pump) $= 472$ ft

6.20   **e)** Given data are $Q_1 = Q_2$, $N_1 = 900$ rev/min. $D_1 : D_2 = 2 : 3$. Use the similitude relation, Eq. 6.3.4 to compute the new pump speed N_2:

$$N_2 = N_1\left(\frac{Q_2}{Q_1}\right)\left(\frac{D_1}{D_2}\right)^3 = 900 \times 1 \times \left(\frac{3}{2}\right)^3 = 3,040 \text{ rev/min}$$

Hydrology

by David A. Hamilton

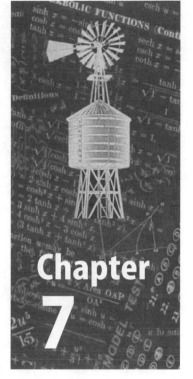

Chapter 7

Solutions to Practice Problems

7.1 b) Use Table 7.2 to compare CN; meadow is consistently less than the other possibilities.

7.2 b) Late winter with bare frozen ground will provide the most rapid and intense response to the rainfall. Almost all of the precipitation will immediately run off. If the deep snow pack melts rapidly with a rainstorm, it would significantly increase the runoff volume and then increase the flood peak. As described, it is likely the snowpack would absorb much of the rainfall, and then hold it with the return of subfreezing temperatures. Late summer has the highest infiltration rate, and more vegetation for interception, so much of the rainfall would not runoff.

7.3 b) There is a 5% chance of this flood flow being equaled or exceeded each year. Over a period of time we would expect it to occur in approximately 5% of the years (i.e., $40 \times 0.05 = 2$).

7.4 e) Base flow tends to be increased because there is greater recharge to groundwater from depressions, so more groundwater tends to be available to sustain base flows. Depressions remove water from surface runoff, which then decreases both volume and flood peak.

7.5 c) Using Eq. 7.7.1: $2 = 100/T$; therefore $T = 50$ years.

Multiple Choice Problems

7.6 **b)**   Answer a) can introduce significant error by neglecting steep reaches with very short travel times; answer c) assumes implicitly average characteristics that may be significantly different form the watershed in question; answer d) assumes the nearby stream has the same T_c and it may be very different.

7.7 **d)**   a)–c) are all integral components of the design. The design storm is chosen based on risk, governmental regulation, or economics. The peak outflow depends on current or projected downstream uses and channel capacity, and governmental regulation. And storage volume is the biggest factor determining the effectiveness of the structure in reducing flood peaks, it is also the major factor in determining the cost for land acquisition and liability associated with deep ponds. Once any two of these factors are determined, the third is automatically fixed.

7.8 **a)**   The effect of storage is to slow down flood peaks. Introducing detention ponds tends to slow down the flood peak, so it reaches downstream areas later. One caution, slowing the timing of the flood peak may cause it to coincide with the peak on a downstream tributary. This has the potential of actually increasing the flood peak below that confluence.

7.9 **b)**   Wells can be used to control the groundwater gradient to contain and capture a plume. Once contamination is in an aquifer, limiting recharge will not contain it. And even if it is cost effective to isolate a plume by artificially constructing a low hydraulic conductivity barrier around it, the gradient must still be controlled or contamination could still slowly move through the barrier.

Essay Problems

7.10 **A)**   First we must consider that there is an additional source of water. The water balance equation must be modified to include I for irrigation as a positive input, along with precipitation. Assume the irrigation system is properly managed so there is no surface runoff or groundwater recharge from precipitation events. Assume the root zone is three feet deep (a reasonable estimate for a problem like this) and on average the soil moisture will be maintained at half its water holding capacity. Assume the soil profile is saturated to its field capacity at the beginning of the season.

From these assumption, $Q = 0$, $G = 0$, and ΔS is the amount of water drawn from storage out of the soil profile. This is half the water holding capacity for a 3 foot thickness of the root zone as assumed. The groundwater storage is

$$\Delta S = 0.5 \times (0.1 \text{ in/in} \times 3 \text{ ft} \times 12 \text{ in/ft}) = 1.8 \text{ in}$$

Inserting I in the water budget equation (7.1.1) allows us to write

$$I = \Delta S + ET - P$$
$$= 1.8 + 24 - 11 = 14.8 \text{ in}$$

The average volume of water needed is:

$$\frac{14.8 \text{ in} \times 5000 \text{ ac}}{12 \text{ in/ft}} = 6167 \text{ ac} - \text{ft}$$

B) Other design considerations are:
- risk analysis: what level of protection against drought is desired
- cost effectiveness
- peak flow demand
- economic and environmental impacts of withdrawing this volume from the source water body
- environmental impacts of irrigation return flow on receiving water bodies
- public acceptance of the project.

7.11 A) From Fig. 7.3, the design rainfall is approximately 4 inches. The *CN* is determined from Table 7.2 to be 80 for the first parcel, and 54 for the alternative site. The runoff is 2.05 in or 17 ac-ft for the first parcel. The runoff is determined from either Fig. 7.5, or Eqs. 7.4.4 and 7.4.5.

B) The runoff for the second site is 0.5 in, or 4 ac-ft. The volume of the detention pond will be significantly less on the A soils.

7.12 The design rainfall is found from Fig. 7.3 as 7.5 in. The *CN* are found in Table 7.2, and the runoff is determined from either Fig. 7.5, or Eqs. 7.4.4 and 7.4.5. The runoff can be calculated for each land use and summed for the watershed total.

Land Use	Soil Type	CN	SRO (in)	A_d (mi^2)	$A_d \times SRO$ (mi^{2}in)
row crop	C	85	5.75	1.5	8.63
1/4 sub	A	61	3.1	0.75	2.33
1/4 sub	B	75	4.6	0.75	3.45
			Total:	3.0	14.4

$$Q_p = q_p \times A_d \times SRO$$

$$= 75 \text{ cfs/mi}^2\text{in} \times 14.4 \text{ mi}^2\text{in} = 1,080 \text{ cfs}$$

7.13 Pipe A–B will carry the combined peak runoff from areas I and II. The longest T_c for the combined area is 7 minutes. Using the IDF equation (Eq. 7.8.3) and the parameters from Table 7.5 for Atlanta, the rainfall intensity is

$$i = \frac{c}{T_d^{\,e} + f}$$

$$= \frac{97.5}{7^{0.83} + 6.88} = 8.19 \text{ in/hr}$$

Therefore, the flow rate is $Q = i \times \sum CA = 8.19 \times (0.75 \times 2 + 0.7 \times 3) = 29.5$ cfs.

Using the ground elevation as a guide for the slope of the pipe,

$$S_0 = (704.7 - 701.0)/500 = 0.0074 \text{ ft / ft}$$

Assume $n = 0.015$ for the pipe. Then from Eq. 7.8.4,

$$D_p = \left(2.16Q\,n/S_0^{0.5}\right)^{3/8}$$

$$= \left(2.16 \times 29.5 \times 0.015/0.0074^{0.5}\right)^{3/8} = 2.47 \text{ ft}$$

Round up to the nearest commercial size, 2.5 ft or 30 in.

Calculate the travel time:

$$V = \frac{Q}{A} = \frac{29.5}{\pi 2.5^2/4} = 6.0 \text{ ft/sec}$$

Therefore, travel time $= \dfrac{L}{V} = 500/6.1 = 83$ sec or 1.4 min.

Pipe B–C will add the runoff from Areas III and IV to pipe A–B. The T_c from Areas I and II plus travel time is $7 + 1.4 = 8.4$ min. Compare this with T_c of 15 min for Area IV. Therefore, 15 min will be used to size Pipe B–C.

All four areas contribute flow. The rainfall intensity is

$$i = \frac{97.5}{15^{0.83} + 6.88} = 5.96 \text{ in/hr.}$$

The flow rate is then $Q = 5.96[3.6 + 0.6 \times 4 + 0.6 \times 5] = 54 \text{ cfs}$.

The slope is $S_0 = \dfrac{701.0 - 698.3}{500} = 0.0054 \text{ ft/ft}$.

And the diameter of the B–C pipe is $D_p = \left(\dfrac{2.16 \times 54 \times 0.015}{0.0054^{0.5}}\right)^{3/8} = 3.28$.

Round up to the nearest commercial size, 3.5 ft or 42 in.

7.14   Use Eq. 7.8.10 to calculate velocity, and Eq. 7.8.11 to calculate the travel time for each reach. Sum all the travel times for the total T_c. The calculations are summarized below. Use $T_c = 6$ hr.

	S (ft/ft)	length (ft)	V (ft/sec)	T_c (hr)
small trib	0.001	5000	.66	2.09
small trib	0.0015	2000	.81	0.68
waterway	0.0007	1500	.32	1.31
small trib	0.0009	3000	.63	1.32
sheet flow	0.003	500	.26	0.53
sheet flow	0.01	200	.48	0.12

Total = 6.05 hr

7.15   Use Eq. 7.8.9 to determine the unit hydrograph. We have

$$\frac{Q_p}{A_d \times SRO} = \frac{484}{T_d/2 + 0.6T_c}$$

$$= \frac{484}{6/2 + 0.6 \times 6} = 73 \text{ cfs}/\text{mi}^2\text{in}$$

Use Fig. 7.5 or Eqs. 7.4.4 and 7.4.5 to find $SRO = 2.05$ in.

Therefore the peak design flood flow from this watershed is

$$Q_p = 73 \frac{\text{cfs}}{\text{mi}^2\text{in}} \times 3 \text{ mi}^2 \times 2.05 \text{ in} = 450 \text{ cfs}$$

7.16   The drainage area is given as 15 square miles. The point rainfall is 5 inches (from Fig. 7.3), which must be adjusted since the drainage area is more than 10 square miles. The point rainfall adjustment from Table 7.1 is 0.978, therefore the design rainfall is 4.89 in. The soils are described as very sandy; assume they are hydrologic soil group A. Use Table 7.2 to determine CN for each land use, and calculate the composite CN for the watershed:

Land Use	Soil Group	CN	Area (mi²)	Product
crops	A	65	9	585
forest	A	45	3	135
1/4 ac sub	A	61	1.5	91.5
wetland	A	85	1.5	127.5
		Totals:	15	939

The composite CN is then $939/15 = 63$, and from Fig. 7.5,

$$SRO = 1.5 \text{ in.}$$

T_L is given as 12 hours, the unit hydrograph peak from Fig. 7.12 is

$$q_p = 47 \text{ cfs}/\text{mi}^2\text{in}$$

The adjustment for 10% wetland throughout the watershed is 0.71 (from Table 7.7). Therefore, the design peak runoff is

$$Q_p = 47 \text{ cfs}/\text{mi}^2\text{in} \times 15 \text{ mi}^2 \times 1.5 \text{ in} \times 0.71 = 750 \text{ cfs}$$

7.17   In Eq. 7.7.1 $p = 1\%$ so that

$$1\% = 100 \times 1/T. \quad \therefore T = 100 \text{ yr}$$

Then using Eq. 7.7.2

$$p(T \text{ event in } n \text{ years}) = 1 - (1 - 1/T)^n \times 100$$

$$= 1 - (1 - 1/100)^{100} \times 100 = 63\%.$$

7.18   First calculate the static water elevations (SWE), ground elevation – depth to water:

Observation well	Ground elev. (ft)	Depth to water (ft)	SWE (ft)
A	925	25	900
B	932	37	895
C	908	18	890
D	927	32	895

A)   Plotting the SWE on the sketch and using triangulation, the flow direction is north.

B)   Rearranging Eq. 7.10.3 and solving for K:

$$K = T/b$$

$$= \frac{10,000 \ ft^2/day}{100 \ ft} = 100 \ ft/day.$$

The gradient can be calculated between A and C since they are in line with the flow direction:

$$I = \frac{10'}{1000'} = 0.01.$$

The total flow through a 1000 foot wide section can be calculated using Eq. 7.10.1:

$$Q = KIA$$

$$= 100 \ ft/day \times 0.01 \times (1000' \times 100') = 100,000 \ ft^3/day.$$

C)   Estimate $n_e = S = 0.25$. Using Eq. 7.10.2 the seepage velocity is

$$V_s = \frac{KI}{n_e} = \frac{100 \ ft/day \times 0.01}{0.25} = 4 \ ft/day.$$

Finally,

$$travel \ time = \frac{distance}{seepage \ velocity} = \frac{1000 \ ft}{4 \ ft/day} = 250 \ days.$$

7.19   Use the SCS triangular unit hydrograph to develop the runoff hydrograph. Note the Rational Formula only estimates peak flows, not the runoff hydrograph. Since the design rainfall is a six hour duration, develop a six hour unit hydrograph. With $T_d = 6$ hr and $T_c = 1$ hr, and since $T_l = 0.6T_c$ and $T_p = T_l + (T_d/2)$ (see Fig. 7.7), therefore

$$T_p = 0.6 \times 1 + (6/2) = 3.6 \ hr.$$

The time of recession is then (see Fig. 7.11)

$$T_r = 1.67 \times T_p = 1.67 \times 3.6 = 6 \ hours.$$

The base of the hydrograph is then 6 + 3.6 = 9.6 hours. From Eq. 7.8.7

$$q_p = \frac{0.75 \times SRO}{T_p}$$

For a unit hydrograph $SRO = 1$ in so that $q_p = 0.75/3.6 = 0.21$ in / hr.

Use Fig. 7.11 as a guide. The runoff from the 6-hour storm (using $CN = 80$, and Fig. 7.5) is $SRO = 2.05$ in. Multiply q_p by this, and 1 square mile and 645 to convert to cfs will yield the maximum runoff rate:

$$Q_p = A_d \times SRO \times q_p$$
$$= 1 \times 2.05 \times 0.21 \times 645 = 275 \text{ cfs.}$$

The hydrograph follows:

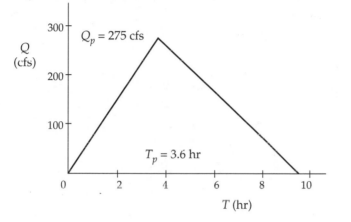

7.20   Other factors to be considered in the design:

1.   Evaluate downstream impacts. The change in flood peak timing may cause an increase in flooding downstream.

2.   Evaluate the impact of extreme flood events on the structure. For example, if the design is to control a 25-year flood peak, what will happen during a 100-year flood? An emergency spillway may be needed to protect the integrity of the structure and downstream people and property.

3.   Have an effective sediment and erosion control program.

4.   Develop operation and maintenance guidelines and schedule. Non-maintenance of detention ponds is the major cause of their failure; make sure maintenance needs are clearly understood, documented and can be done.

5.   Determine cost effectiveness.

Structural Steel

by Richard W. Furlong

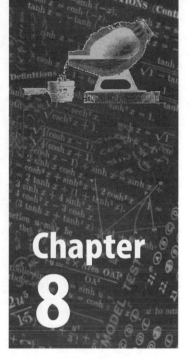

Chapter 8

Solutions to Practice Problems

8.1 **LRFD**. *Step 1*—Compute the tensile strength of the strap.

$$T_u = \phi A_g F_y = 0.90\left(6 \times \tfrac{5}{8}\right)36 = 122 \text{ k}$$

Step 2—Select weld size and electrode. **LRFD** Table J2.5 requires at least 1/4 in weld size for 5/8 in thick plate. Try 5/16 in fillet weld. Since F_u for A36 is 58 ksi, use electrode E60.

Step 3—Compute weld value using Eq. 8.6.1a.

$$W_u = 0.325 \times (5/16) \times 60 = 6.1 \text{ k/in}$$

Step 4—Compute required length of weld = $T_u/W_u = 122/6.1 = 20$ in.

Step 5—Sketch connection:

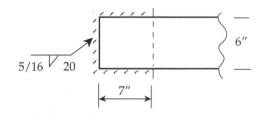

5/16 ⟍ 20

7″

6″

LRFD. *Step 1*—Compute allowable force $T = 0.6F_y A_g$. $T = 0.6 \times 36 \times 6 \times \dfrac{5}{8} = 81$ k.

Step 2—Select weld size and electrode quality. **LRFD** Table J2.4, pg 5-67 requires electrode at least 1/4 in size for plate thickness from 1/2 in to 3/4 in. Select 5/16 in fillet weld. Since F_u for A36 steel is 58 ksi; use E60 electrode.

Step 3—Use Eq. 8.6.1b to compute a weld value $W = 0.18D_wE_{xx} = 0.18(5/16)60 = 3.37$ k/in.

Step 4—Compute required length of weld $= T /W = 81/3.37 = 24$ in.

Step 5—Sketch connection:

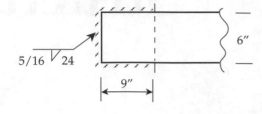

8.2   **LRFD**. *Step 1*—Compute required

$$P_u = 1.2D + 1.6L = 1.2(32) + 1.6(58) = 131 \text{ k}$$

Step 2—Use Eq. 8.6.1a to determine weld value

$$W_u = 0.325D_wE_{xx} = 0.325(1/4)70 = 5.7 \text{ k/in.}$$

Step 3—Compute length of weld $= \dfrac{P_u}{W_u} = \dfrac{131}{5.7} = 23$ in. Use $11\frac{1}{2}$ in of weld on each angle, $2\frac{1}{2}$ in at end, 6 in along heel, and 3 in along toe.

LRFD. *Step 1*—Compute required service load $= D + L = 32 + 58 = 90$ k.

Step 2—Use Eq. 8.6.1b to determine weld value

$$W = 0.18D_wE_{xx} = 0.18(1/4)70 = 3.15 \text{ k/in.}$$

Step 3—Compute length of weld $= \dfrac{P}{W} = \dfrac{90}{3.15} = 29$ in. Use $14\frac{1}{2}$ in of weld on each angle, $2\frac{1}{2}$ in at end, 8 in along heel, and 3 in along toe.

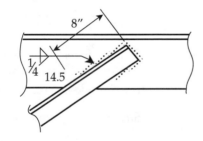

8.3   **LRFD**. *Step 1*—Bolts are in double shear. Use Table 1-D, pg 5-5: A307—$T_u = 14.3 \times 3 = 42.9$ k and A325—$T_u = 41.4 \times 3 = 124.2$ k.

Step 2—Determine bolt bearing strength (pg 1-30) for W14 × 68, $t_w = 0.415$ in. Use Table 1-E (pg 5-7) for 1 in thick $R_b = 91.3$ k. For $t_w = 0.415(91.3)3 = 113.7$ k for both bolts. The thickness of the 2 angles exceeds that of the beam web, so bearing is limited by beam web.

Step 3—Check block shear (independent of bolt quality). Dimensions of shear block are identical for angles and for beam web. The thinner web will restrain the smaller force. Tension yield and shear fracture:

$$R_{BS} = (1.50)0.415(36) + (7.50 - 2.5 \times 7/8)0.415(0.6)58 = 99 \text{ k}$$

Tension fracture and shear yield:

$$R_{BS} = \left(1.50 - \frac{1}{2} \times \frac{7}{8}\right)0.415(58) + 7.50(0.415)(0.6)36 = 93 \text{ k}$$

The <u>larger</u> value governs.

Step 4—Report the least value as connection strength.

A307 Bolts $T_u = 42.9$ k, governed by bolt shear.

A325X Bolts $T_u = 99.0$ k, governed by block shear.

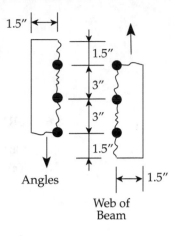

1.5"

1.5"

3"

3"

1.5"

Angles   1.5"

Web of
Beam

8.4 **LRFD**. *Step 1*—Use Table 1-A, pg 5-3. For A325 bolt 7/8 in diameter, $T_u = 40.6$ k/bolt. For 4 bolts:

$$T_u = 4(40.6) = 163 \text{ k}$$

LRFD. *Step 1*—Use Table 1-A, pg 4-3. For A325 bolt 7/8 in diameter, $T = 26.5$ k/bolt. For 4 bolts:

$$T = 4(26.5) = 103 \text{ k}$$

Allowable

8.5 **LRFD**. *Step 1*—Use equations of Table J3-3, pg 6-68. Compute:

$$f_v = \frac{P_u \cos \alpha}{A_b} = \frac{0.8 P_u}{6(0.44)} = 0.303 P_u$$

$$f_t = \frac{P_u \sin \alpha}{A_b} = \frac{0.6 P_u}{6(0.44)} = 0.227 P_u$$

Solve governing equation for P_u:

$$f_t = 85 - 1.4 f_v \leq 68 \text{ ksi}$$

$$0.227 P_u = 85 - 1.4(0.303 P_u)$$

$$P_u = 130 \text{ k}$$

Step 2—Check $f_t = 0.227 P_u = 0.227(130) = 30$ ksi < 68 ksi. $\therefore$ OK.

LRFD. *Step 1*—Use equations of Table J3-3, pg 5-74.

$$f_v = \frac{P \cos \alpha}{A_b} = \frac{0.8 P}{6(0.44)} = 0.303 P$$

$$f_t = \frac{P \sin \alpha}{A_b} = \frac{0.6 P}{6(0.44)} = 0.227 P$$

Solve governing equation for P:

$$f_t = \sqrt{44^2 - 2.15f_v^2}$$

$$(0.303P)^2 = 44^2 - 2.15(0.227P)^2$$

$$P = 97.8 \text{ k}$$

8.6   **LRFD.** *Step 1*—Use Clause J3.5, pg 6-68. Permissible shear $= R_v\left(1 - \dfrac{T}{T_b}\right)$.

Obtain $R_v = 17A_b$ from Table J3.4, pg 6-69. Obtain $T_b = 39$ k/bolt from Table J3.1, pg 6-66. Then,

$$T\cos\alpha = 17A_b\left(1 - \frac{T\sin\alpha}{39n}\right)$$

where n = number of bolts. Solve for T:

$$T = 17(6 \times 0.44)\left(1 - \frac{T \times 0.6}{39 \times 6}\right)\bigg/ 0.8$$

$$= 49 \text{ k}$$

LRFD. *Step 1*—Use Clause J3-6, pg 5-74. From Table J3.2, $f_v = 17$ ksi, and from Table J3.7, $T_b = 28$ k/bolt.

$$\frac{1}{A_b}T\cos\alpha = f_v\left(1 - f_t A_b / T_b\right)$$

$$\frac{T(0.8)}{0.44(6)} = 17\left(1 - \frac{T(0.6)}{A_b}A_b \times \frac{1}{6 \times 28}\right)$$

$$T = 47 \text{ k}$$

8.7   *Step 1*—Determine properties of bolt group:

$$n = 6 \text{ bolts} = A$$

$$I_x = \sum Ay^2 = 4(3 \text{ in})^2 = 36 \text{ bolt} \cdot \text{in}^2$$

$$I_y = \sum Ax^2 = 6(2.75 \text{ in})^2 = 45.4 \text{ bolt} \cdot \text{in}^2$$

$$\therefore J = I_x + I_y = 36 + 45.4 = 81.4 \text{ bolt} \cdot \text{in}^2$$

Step 2—Determine vertical and horizontal forces on each bolt, using

$$F_y = \frac{P_y}{A} + \frac{Pe_y}{J} = \frac{31.5}{6} + \frac{31.5(9)(\pm 2.75)}{81.4} = 14.8 \text{ k/bolt max}$$

$$Fx = \frac{P_x}{A} \pm \frac{Pe_x}{J} = 0 + \frac{31.5(9)(\pm 3)}{81.4} = 10.5 \text{ k/bolt max}$$

Step 3—Bolts closest to line of force and farthest from centroid of bolts will always resist highest force. Bolts at upper right and lower right corners have maximum force. For bolt on upper right:

$$F_{\max} = \sqrt{F_x^2 + F_y^2} = \sqrt{14.8^2 + 10.5^2} = 18.1 \text{ k/bolt}$$

8.8   *Step 1*—Determine properties of bolt group (exactly the same as *Step 1* in Problem 8.7):

$$I_x = 36 \text{ bolt} \cdot \text{in}^2, \quad I_y = 45.4 \text{ bolt} \cdot \text{in}^2, \quad J = 81.4 \text{ bolt} \cdot \text{in}^2$$

Step 2—Determine the moment on the connection:

$$M = P_y e_x + P_x e_y = 31.5(9) + 12(6) = 355 \text{ in} \cdot \text{k}$$

Step 3—Determine vertical and horizontal forces on bolt:

$$F_x = \frac{P_y}{A} + \frac{My}{J} = \frac{31.5}{6} + \frac{355(\pm 2.75)}{81.4} = 17.25 \text{ k/bolt max}$$

$$F_x = \frac{P_x}{A} + \frac{Mx}{J} = \frac{12.0}{6} + \frac{355(\pm 3)}{81.4} = 15.1 \text{ k/bolt max}$$

Step 4—The maximum force per bolt will act on the top right bolt, the bolt nearest the line of action of the applied force:

$$F_{max} = \sqrt{F_x^2 + F_y^2} = \sqrt{17.25^2 + 15.1^2} = 22.9 \text{ k/bolt}$$

8.9   *Step 1*—Determine properties of the welds:

$$A = \text{length of weld} = 2(8+10) = 36 \text{ in}$$

$$I_x = \sum Ay^2 + I_{ox} = 2(6)(5)^2 + \frac{2(1)(10)^3}{12} = 467 \text{ weld} \cdot \text{in}^3$$

$$I_y = \sum Ax^2 + I_{oy} = 2(10)(4)^2 + \frac{2(1)(6)^3}{12} = 356 \text{ weld} \cdot \text{in}^3$$

$$\therefore J = 467 + 356 = 823 \text{ weld} \cdot \text{in}^3$$

Step 2—The portion of weld nearest the line of the force at the top edge is the most highly stressed:

$$M = Pe_x = 31.5(9) = 284 \text{ in} \cdot \text{k}$$

$$F_y = \frac{P_y}{A} + \frac{Mc_x}{J} = \frac{31.5}{36} + \frac{284}{823} = 2.25 \text{ k/in}$$

$$F_x = \frac{P_x}{A} + \frac{Mc_y}{J} = 0 + \frac{284(5)}{823} = 1.72 \text{ k/in}$$

$$\therefore F_{max} = \sqrt{2.25^2 + 1.72^2} = 2.83 \text{ k/in}$$

8.10 *Step 1*—Determine properties of the welds. The same welds as Problem 8.9:

$$A = \text{length of weld} = 36 \text{ in}$$

$$I_x = 467 \text{ weld} \cdot \text{in}^3, \quad I_y = 356 \text{ weld} \cdot \text{in}^3$$

$$\therefore J = 823 \text{ weld} \cdot \text{in}^3$$

Step 2—Determine the vertical and horizontal force components at the top edge closest to the line of action of the force:

$$M = P_y e_x + P_x e_y = 31.5(9) + 12(5) = 344 \text{ in} \cdot \text{k}$$

$$F_y = \frac{31.5}{36} + \frac{344(4)}{823} = 2.54 \text{ k/in}$$

$$F_x = \frac{12}{36} + \frac{344(5)}{823} = 2.42 \text{ k/in}$$

$$\therefore F_{max} = \sqrt{2.54^2 + 2.42^2} = 3.51 \text{ k/in}$$

8.11   **LRFD**. *Step 1*—Compute ultimate reaction

$$P_u = 1.2 P_D + 1.6 P_L = 1.2(28) + 1.6(40) = 98 \text{ k}.$$

Ultimate bearing pressure $= \dfrac{98}{7 \times 17} = 0.82$ psi. At toe of flange,

$$M_u = 7(0.82)\left(\frac{17}{2} - 1.31\right)^2 \frac{1}{2} = 148 \text{ in} \cdot \text{k}$$

Step 2—Determine plate thickness. Make plate thickness t large enough that $\phi M_n > 148$ in·k:

$$\phi M_n = \phi F_y \frac{bt^2}{6}$$

$$148 \leq 0.9(36)\frac{(7)t^2}{6}$$

$$\therefore t \geq 1.98$$

Make $t = 2$ in. Choose a bearing plate 7 in $\times$ 2 in $\times$ 1 ft-5 in.

LRFD. *Step 1*—Determine plate thickness such that maximum bending stress is less than $0.75F_y$. Compute bearing stress $= \dfrac{P}{A_R} = \dfrac{28 + 40}{7 \times 17} = 0.57$ ksi.

Max plate M at k from $\mathbf{\mathcal{C}} = 0.57(7)\left(\dfrac{17}{2} - 1.31\right)^2 \dfrac{1}{2} = 103$ in·k.

$$0.75 F_y > \frac{6M}{bt^2}$$

$$t > \sqrt{\frac{6 \times 103}{7(0.75)36}} = 1.80 \text{ in}$$

Make $t = 1\frac{7}{8}$. Bearing plate $7 \times 1\frac{7}{8} \times 1$ ft - 5 in.

8.12   **LRFD**. Refer to Fig. 8.20(b) for plate sized to cover the column section. With plate 10 in $\times$ 14 in. *Step 1*—Compute limit bearing pressure:

$$F_p = 1.7 f'_c = 1.7(30) = 5.1 \text{ ksi}$$

Step 2—Compute

$$P_u = 1.2 P_D + 1.6 P_L$$

$$= 1.2(72) + 1.6(66) = 192 \text{ k}$$

Step 3—Solve for L, Fig. 8.20(b), for 10 in $\times$ 14 in plate:

$$A_{PL} = 2BL + 2L(H - 2L)$$

$$\frac{192}{5.1} = 2(10)L + 2L(14 - 2L)$$

$$4L^2 - 48L = 37.6$$

$$L^2 - 12L + 6^2 = -9.4 + 36$$

$$L = 6 - \sqrt{26.6} = 0.84 \text{ in}$$

Step 4—Solve for $t_p = \sqrt{\dfrac{3F_p L^2}{\phi F_y}} = \sqrt{\dfrac{3(5.1)0.84^2}{0.9(36)}} = 0.63$ in. Base plate

$10 \times \dfrac{3}{4} \times 1$ ft - 2 in.

LRFD. Refer to Fig. 8.20(b) for a plate sized to cover the column section with 10 in $\times$ 14 in plate.

Step 1—Compute allowable bearing pressure $F_p = 0.7 f_c' = 2.1$ ksi.

$$A_{PL} = \frac{P + L}{F_p} = \frac{(72 + 66)}{2.1} = 66 \text{ in}^2$$

Step 2—Solve for L, Fig. 8.20(b).

$$A_{PL} = 2BL + 2L(H - 2L)$$

$$66 = 2(10)L + 2L(14 - 2L)$$

$$66 = 48L - 4L^2$$

$$L^2 - 12L + 6^2 = -16.5 + 36$$

$$L = 6 - \sqrt{19.5} = 1.58 \text{ in}$$

Step 3—Solve for $t_p = \sqrt{\dfrac{3F_p L^2}{0.66 F_y}} = \sqrt{\dfrac{3(2.1)1.58^2}{24}} = 0.81$ in. Base plate

$10 \times \dfrac{7}{8} \times 1$ ft - 2 in.

8.13 **LRFD.** *Step 1*—Compute required force

$P_u = 1.2D + 1.6L = 1.2(227) + 1.6(185) = 568$ k.

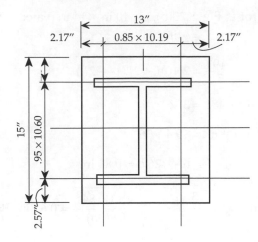

Step 2—Determine required plate size, A_{PL}.

$$A_c = \text{Area top of pier} = \frac{\pi}{4}(30)^2 = 707 \text{ in}^2$$

Assume $F_p = 1.7f_c'$ and find $A_{PL} = \frac{P_u}{\phi F_p} = \frac{568}{0.6(1.7 \times 3)} = 186 \text{ in}^2$. Try 13×15,

$A_{PL} = 195 \text{ in}^2$. Use Eq. 8.6.9a to check A_{PL}:

$$\phi F_p A_{PL} = \phi A_{PL}\left(0.85f_c'\right)\sqrt{\frac{A_c}{A_{PL}}}$$

$$= 0.6(195)(0.85 \times 3)\sqrt{\frac{707}{195}} = 568 \text{ k}$$

This is exactly enough.

Step 3—Use section dimensions to sketch base plate and to find longer distance = 2.57":

$$f_p = \frac{P_u}{A_{PL}} = \frac{568}{195} = 2.91 \text{ ksi}$$

Step 4—Determine thickness such that plate will not yield at cantilever:

$$\phi F_y > 3f_p \, m^2/t_p^2$$

$$0.9(36) > 3(2.91)(2.57)^2/t_p^2$$

$$t_p > 0.77$$

Use plate $13 \text{ in} \times \frac{7}{8} \text{ in} \times 1 \text{ ft-3 in}$.

LRFD. *Step 1*—Determine A_{PL}. Compute $A_c = \frac{\pi}{4}(30)^2 = 707 \text{ in}^2$. Assume allowable $f_p = 0.7f_c'$ and solve for A_{PL}:

$$0.7f_c' A_{PL} = P$$

$$A_{PL} = \frac{(227 + 185)}{0.7(3)} = 197 \text{ in}^2$$

Check Eq. 8.6.9b,

$$A_{PL}\left[0.35f_c'\sqrt{\frac{A_c}{A_{PL}}}\right] \geq P$$

$$224\left[0.35(3)\sqrt{\frac{707}{224}}\right] = 418\ \text{k} > (227 + 185)$$

So, this is acceptable.

Step 2—Use section dimensions to establish longer cantilever distance = 3.07 in.

Step 3—Determine t_p such that plate will not be overstressed at 3.07 in cantilever:

$$0.66F_y > 3f_p m^2/t_p^2$$

$$27 > 3\left(\frac{412}{224}\right)3.07^2/t_p^2$$

$$t_p = 1.39\ \text{in}$$

Use plate 14 in $\times 1\frac{1}{2}$ in $\times 1$ ft - 4 in.

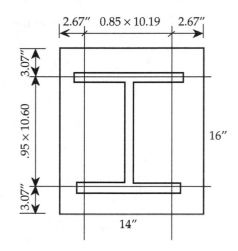

8.14 **(b)**   **LRFD**

$$P_{\text{live}} = (20/16)50 = 62.5\ \text{kip}$$

$$P_u = 1.2P_{\text{dead}} + 1.6P_{\text{live}} = 1.6 \times 62.5 = 100\ \text{kip}$$

$$\psi F_u A_e \geq P_u;\ A_e = UA_n$$

$$A_g = A_n > P_u/(\phi F_u U) = 100/(0.75 \times 65 \times 0.85) = 2.41\ \text{in}^2$$

Try $2L2 \times 2 \times 3/8(A_g = 2.72)$; verify that $\psi F_y A_g \geq P_u$

ASD

$$P = (20/16)50 = 62.5\ \text{kip}$$

$$f_t \leq 0.5F_u;\ f_t = P/(UA_n)$$

$$A_{g\ \text{req'd}} = A_{n\ \text{req'd}} = P/(0.5F_u U)$$

$$= 62.5/(0.5 \times 65 \times 0.85) = 2.26\ \text{in}^2$$

Try $2L2 \times 2 \times 5/16(A_g = 2.30)$; verify that $f_t = P/A_g \leq 0.6F_y$

Multiple Choice Problems

8.15 (b)   LRFD

$$P_{\text{dead}} = w_{\text{dead}}L/2 = 17.5 \times 16/2 = 140 \text{ kip}$$

$$\begin{aligned} P_{\text{live}} &= (w_{\text{live}}L/2) + (12/16)P_{\text{live}} \\ &= (5 \times 16/2) + (12/16)\,50 = 77.5 \text{ kip} \end{aligned}$$

$$P_u = 1.2 \times 140 + 1.6 \times 77.5 = 292 \text{ kip}$$

$$K_y L_y = 1.0 \times 12 = 12 \text{ ft}$$

$W8 \times 35$ from column design tables

ASD

$$P = wL/2 + (12/16)\,P_{\text{live}}$$

$$= (17.5 + 5)\,16/2 + (12.16)\,50 = 218 \text{ kip}$$

$$K_y L_y = 1.0 \times 12 = 12 \text{ ft}$$

$W10 \times 39$ from column design tables

8.16 (a)   LRFD

$$P_u = 292 \text{ kip}$$

Assume $r_x/r_y = 2$

$$(K_y L_y)_{\text{eff}} = K_x L_x/(r_x/r_y) = 12/2 = 6 \text{ ft}$$

Try $W8 \times 28$ from column design tables. $r_x/r_y = 2.13$

$$(K_y L_y)_{\text{eff}} = K_x L_x/(r_x/r_y) = 12/2.13 = 5.63 \text{ ft}$$

$$\phi P_n = 306 \text{ kip} \geq 292. \quad \therefore \text{OK}.$$

ASD

$$P = 218 \text{ kip}$$

Assume $r_x/r_y = 2$

$$(K_y L_y)_{\text{eff}} = K_x L_x/(r_x/r_y) = 12/2 = 6 \text{ ft}$$

Try $W8 \times 31$ from column design tables. $r_x/r_y = 1.72$

$$(K_y L_y)_{\text{eff}} = K_x L_x/(r_x/r_y) = 12/1.72 = 6.98 \text{ ft}$$

$$\text{capacity} = 234 \text{ kip} \geq 180. \quad \therefore \text{OK}.$$

8.17 (c)   LRFD

$$M_{\text{dead}} = wL^2/8 = 17.5 \times 16^2/8 = 560 \text{ kip-ft}$$

$$M_{\text{live}} = 5 \times 16^2/8 = 160 \text{ kip-ft}$$

$$M_u = 1.2 \times 560 + 1.6 \times 160 = 928 \text{ kip-ft}$$

$W30 \times 90$ from Z_x economy table

ASD

$w = 5 + 17.5 = 22.5 \text{ kip-ft}$

$M_{\text{req'd}} = mL^2 / 8 = 22.5 \times 16^2 / 8 = 720 \text{ kip-ft}$

$W30 \times 99$ from S_x economy table

8.18 (d) LRFD

$M_u = 928 \text{ kip-ft}$

$L_b = 16 \text{ ft}; \ C_b = 1.14$

$M_u / C_b = 928 / 1.14 = 814 \text{ kip-ft}$

$W30 \times 90$ from beam design charts

$\phi M_{cr} = 827 \geq 814; \ \phi M_p = 1060 \geq 928$

ASD

$M_{\text{req'd}} = 720 \text{ kip-ft}$

unbraced length $= 16 \text{ ft}; \ C_b = 1.0$

$W33 \times 118$ from beam design charts

8.19 (a) LRFD

$h / t_w \leq 418 / \sqrt{F_y}$? $43.9 \leq 59.1.$ Yes, so

$\phi V_n = \phi 0.6 F_y A_w = \phi 0.6 F_y d t_w$

$\qquad = 0.9 \times 0.6 \times 50 \times 30.31 \times 0.615 = 503 \text{ kip}$

ASD

$h / t_w \leq 380 / \sqrt{F_y}$? $43.9 \leq 53.7.$ Yes, so

$f_v = R / (d t_w) = 180 / (30.31 \times 0.615) = 9.7 \text{ ksi}$

8.20 (e) LRFD

$P_{\text{dead}} = w_{\text{dead}} L / 2 = 17.5 \times 16 / 2 = 140 \text{ kip}$

$P_{\text{live}} = w_{\text{live}} L / 2 = 5 \times 16 / 2 = 40 \text{ kip}$

$P_u = 1.2 P_{\text{dead}} + 1.6 P_{\text{live}} = 1.2 \times 140 + 1.6 \times 40 = 232 \text{ kip}$

$M_{\text{ntx}} = P_u e = 232 \times 7 / 12 = 135 \text{ kip-ft}$

ASD

$P = wL / 2 = (17.5 + 5)16 / 2 = 180 \text{ kip}$

$f_{bx} = M / S_x = P e / S_x = 180 \times 7 / 35.5 = 35.5 \text{ ksi}$

8.21 (c) **LRFD**

$\phi F_v A_b = 43.3$ kip/bolt; $R_u = 232$ kip

$272 / 43.3 = 5.4$ so use 6 rows (also verify $n\phi 2.4 F_u d t \geq V_u$)

3 lines $\times$ 6 rows $=$ 18 bolts

ASD

capacity $= 25.3$ kip/bolt;   required strength $= 180$ kip

$180/25.3 = 7.1$ so use 8 rows (also verify $V/ndt \leq 1.2 F_u$)

3 lines $\times$ 8 rows $=$ 24 bolts

8.22 (d) **LRFD**

$A_{nv} = 2 \times 3/8 \times \left[21 - 7 \times (7/8 + 1/16 + 1/16)\right] = 10.5$ in^2

$\phi 0.6 F_u A_{nv} = 0.75 \times 0.6 \times 65 \times 10.5 = 307$ kip

ASD

$A_{nv} = 2 \times 3/8 \times \left[21 - 7 \times (7/8 + 1/16 + 1/16)\right] = 10.5$ in^2

$f_v = R / A_{nv} = 180 / 10.5 = 17.1$ ksi

8.23 (c) **LRFD**

$P_u = 100$ kip

design strength per inch of weld

$\quad$ = lesser of weld shear strength or base metal shear strength

$\phi 0.6 F_{Exx} A_w / l_w = 0.75 \times 0.60 \times 70 \times (.707 \times 3/16) = 4.18$ kip/in

$\phi 0.6 F_u A_{BM} / l_w = 0.75 \times 0.6 \times 65 \times 1/4 = 7.31$ kip/in

$100/4.18 = 23.9$ inches

ASD

$P = 62.5$ kip

capacity per inch of weld

$\quad$ = lesser of weld shear capacity or base metal shear capacity

$0.3 F_{Exx} A_w / l_w = 0.3 \times 70 \times (.707 \times 3/16) = 2.78$ kip/in

$0.3 F_u A_{BM} / l_w = 0.3 \times 65 \times 1/4 = 4.88$ kip/in

$62.5/2.78 = 22.5$ inches

Reinforced Concrete

by Richard W. Furlong

Chapter 9

Solutions to Practice Problems

Multiple Choice Problems

9.1 **b)** *Step 1*—Compute factored load:

$$1.4 \times w_d = 1.4 \times 0.75 = 1.05 \text{ k/ft}$$
$$1.7 \times w_d = 1.7 \times 1.30 = 2.21$$
$$\overline{w_u = 3.26 \text{ k/ft}}$$

$$P_u = 1.7 \times 25 = 42.5 \text{ k}$$

Step 2—Compute moment at mid-span:

$$M_u = \frac{1}{8} w_u \ell^2 + \frac{P_u \ell}{4}$$
$$= \frac{1}{8} \times 3.26 \times 26^2 + 42.5 \times \frac{26}{4} = 552 \text{ ft-k}$$

9.2 **b)** *Step 1*—Compute $A_s = 4 \times (1.00 + 0.79) = 7.16 \text{ in.}^2$

Step 2— Compute $a = \dfrac{A_s f_y}{0.85 f'_c b} = \dfrac{7.16 \times 60}{0.85 \times 4 \times 78} = 1.62$

Step 3 — Moment capacity $= \phi M_n = 0.90 \times A_s f_y \left(d - \dfrac{a}{2} \right)$

$$= 0.90 \times 7.16 \times 60 \left(18 - \dfrac{1.62}{2} \right) \times \dfrac{1}{12} = 554 \ \text{ft-k}$$

9.3 d) *Step 1*—Determine the critical shear section at a distance d from face of support (wall). Since face of wall is $12/2 = 6''$ from center of support, the critical shear section from the center of support is $d + 6'' = 18 + 6 = 24''$.

Step 2—Reaction $= \dfrac{42.5 + 26 \times 3.26}{2} = 63.6 \ \text{k}$

Step 3—Shear at critical section: $v_u = 63.6 - 3.26 \times \dfrac{24}{12} = 57.1 \ \text{k}$

9.4 c) *Step 1*—Maximum spacing for #3 stirrups:

$$s_{max} = \text{Min.} \left(\dfrac{d}{2} = 9'', \ 24'', \ A_v f_y / 50 b_w \right)$$

$$\dfrac{A_v f_y}{50 b_w} = \dfrac{0.22 \times 60 \ 000}{50 \times 12} = 22''$$

$$\therefore s_{max} = 9''$$

Step 2—

$$\phi V_c = 0.85 \times 2 \sqrt{f'_c} \ bd$$

$$= 0.85 \times 2 \sqrt{4000} \times 12 \times 18 = 23,200 \ \text{lb} = 23.2 \ \text{k}$$

Step 3—

$$s = \dfrac{\phi A_v f_y d}{V_u - \phi V_c} = \dfrac{0.85 \times 0.22 \times 60 \times 18}{57.1 - 23.2} = 6'' \ \left(< s_{max} = 9'' \right)$$

$$\therefore \ \text{Use } 6''$$

9.5 e) *Step 1*—Calculate strength at section with only the bottom layer of bars:

$$0.9 M_n = 0.9 A_s \times f_y \left(d - \dfrac{a}{2} \right)$$

$$A_s = 3.57 \ \text{m}^2$$

$$a = \dfrac{3.57 \times 60}{0.85 \times 4 \times 78} = 0.81''$$

$$0.9 M_n = 0.9 \times 3.57 \times 60 \left[19.0 - \left(0.81/2 \right) \right] \div 12 = 299 \ \text{ft-k}$$

Step 2—Find x = distance from support, where moment equals $0.9 M_n$:

$$299 = 63.6x - \dfrac{3.26}{2} x^2$$

$$\therefore x = 5.47 \ \text{ft}$$

Step 3—Development length of #9 (for clear spacing for 2 bar diameters):

$$\ell_d = 2 \times 3.17 \ \text{ft} = 6.34 \ \text{ft}$$

Step 4—Cut-off must extend over max. (1.5 ft, 12 d_b = 1.5 ft

Step 5—Cut-off point from support = 5.47 ft – 1.5 ft ≅ 4 ft

Step 6—Length of bars = 26′ – 2 × 4 = 18 ft

9.6 **a)** *Step 1*—Find c_b at concrete strain = 0.003 = ε_u

and steel strain = ε_y = 60 / 29,000 = 0.00207.

$d = 20'' - 2.5'' = 17.5''$

$$c_b = 17.5 \times \frac{\varepsilon_u}{\varepsilon_u + \varepsilon_y} = 17.5 \frac{0.003}{0.00507} = 10.36''$$

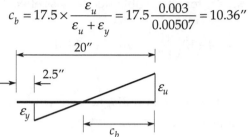

Step 2—Find the strain in the compression steel:

$$= 0.003 \times \frac{10.36 - 2.5}{10.36} = 0.0023 > \varepsilon_y$$

∴ compression steel yielded $f's$ = 60 ksi

Step 3—Axial Force Resistance:

$$P_b = C + A'_s f_y - A_s f_y$$
$$= 0.85 \times f'_c \times b \times a + \left(A'_s - A_s\right) f_y$$
$$= 0.85 \times 3.5 \times 12 \times (0.85 \times 10.36) - 0 \times f_y$$
$$= 314 \text{ k}$$

9.7 **c)** The moment resistance is obtained by taking moments of the axial forces about the center line:

$$M_b = 0.85 \times f'_c \times b \times a \left(\frac{h}{2} - \frac{a}{2}\right) + A'_s f'_s \left(\frac{h}{2} - d'\right) + A_s f_s \left(d - \frac{h}{2}\right)$$
$$= 0.85 \times 3.5 \times 8.8 \times 12 \left(\frac{20}{2} - \frac{8.8}{2}\right) + 2 \times 60 \left(\frac{20}{2} - 2.5\right)$$
$$+ 2 \times 60 \left(17.5 - \frac{20}{2}\right) = 296.6 \text{ ft-k}$$

$$\left(\text{eccentricity} = e_b = \frac{M_b}{P_b} = 11.3 \text{ in.}\right)$$

9.8 **e)** Axial Load Capacity (e = 0.0):

$$P_n = 0.85 \times 3.5 \times 12 \times 20 + 4 \times 60 = 954 \text{ k}$$

9.9 **a)** Maximum spacing = minimum of:

(a) $48 \times 3/8'' = 18''$

(b) $16 \times 9/8'' = 18''$

(c) $12''$ ← governs

9.10 **d)** *Step 1*—For braced frame use $k = 1.0$ radius of gyration $= 0.3\,h = 6''$

Step 2 — Max. $\dfrac{k\ell}{r}$ for "negligible slenderness effect" $= 34 - 12\left(\dfrac{M_1}{M_2}\right)$

$$\dfrac{1 \times \ell}{6} = 34 - 12\left(\dfrac{60}{120}\right) = 28. \quad \ell = 14 \text{ ft}$$

9.11 **b)** *Step 1*—Consider the interaction diagram shown. Assume linear variation between points $\left(0, P_o\right)$ and $\left(M_b, P_b\right)$

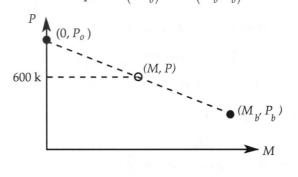

$$P_o = 954, \quad P_b = 314, \quad M_b = 297$$

Step 2—For $P = 600$ k,

$$M = \left(954 - 600\right)\dfrac{297}{954 - 314} = 164 \text{ ft - k}$$

$$\therefore e = \dfrac{164}{600} \times 12 = 3.3 \text{ in}$$

Indeterminate Structures

by Ronald S. Harichandran

Chapter 10

Solutions to Problems

Essay Problems

Question 10.1

1. Determine loads. For maximum deflection, place uniform live load over entire span and concentrated live load at U_4.

    Loads at U_1, U_2, U_3, U_5, U_6 and U_7 due to uniform live load + impact= 0.8(25)(1.15) = 23 kips

    Loads at U_0 and U_8 due to uniform live load + impact=23/2 = 11.5 kips

    Load at U_4 due to uniform live load + concentrated live load + impact= 23 + 18(1.15)

    $$= 43.7 \text{ kips}$$

2. Tabulate bar forces due to live load P_i and due to unit load p_i applied at U_4, and compute $P_i p_i L_i / A_i$ for bars on the left half of the truss.

Bar	L (in.)	A (in^2)	P (kips)	p (kips)	PpL/A
$U_0 U_1$	300	17.9	0	0	0
$U_1 U_2$	300	51.8	−158.7	−1	919.11
$U_2 U_3$	300	51.8	−158.7	−1	919.11
$U_3 U_4$	300	68.5	−225.4	−2	1974.31
$L_0 L_1$	300	26.5	90.85	0.5	514.25
$L_1 L_2$	300	26.5	90.85	0.5	514.25
$L_2 L_3$	300	51.8	137.7	1.5	1178.86
$L_3 L_4$	300	51.8	137.7	1.5	1178.86
$U_0 L_0$	300	17.9	−11.5	0	0
$U_1 L_1$	300	8.85	0	0	0
$U_2 L_2$	300	17.9	−23	0	0
$U_3 L_3$	300	8.85	0	0	0
$U_4 L_4$	300	17.9	−43.7	−1	732.40
$L_0 U_1$	424.26	51.8	−128.48	−0.7071	744.08
$U_1 L_2$	424.26	26.5	95.95	0.7071	1086.21
$L_2 U_3$	424.26	29.1	−63.43	−0.7071	653.90
$U_3 L_4$	424.26	17.9	30.9	0.7071	517.87
					$\Sigma = 10968$

All members except for $U_4 L_4$ have mirror images on the right side of the truss and must therefore be included twice in the displacement calculation. The vertical deflection at U_4 is

$$\Delta = \frac{1}{E}\sum_i \frac{P_i p_i L_i}{A_i} = \frac{1}{29000}[2(10968) - 732.4] = 0.73 \text{ in.}$$

Question 10.2

1. Determine loads. For maximum deflection apply the uniform load over the entire span and the concentrated load at mid-span.

    Uniform live load + impact $= 0.4(1.25) = 0.5$ kips/ft

    Concentrated live load + impact $= 11.5(1.25) = 14.375$ kips

2. Determine the bending moment variations due to the live load and a unit load at mid-span.

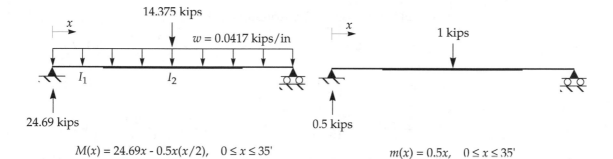

$$M(x) = 24.69x - 0.5x(x/2), \quad 0 \le x \le 35'$$

$$m(x) = 0.5x, \quad 0 \le x \le 35'$$

3. Calculate the midspan deflection.

    $$E = 29000(144) = 4.176 \times 10^6 \text{ k/ft}^2$$

    $$I_1 - 18,300/12^4 - 0.8825 \text{ ft}^4 \qquad I_2 = 27,200/12^4 = 1.3117 \text{ ft}^4$$

    $$\Delta = \int \frac{mM}{EI} dx = \frac{2}{EI_1} \int_0^{15'} 0.5x(24.69x - 0.25x^2)dx + \frac{2}{EI_2} \int_{15'}^{35'} 0.5x(24.69x - 0.25x^2)dx$$

    $$= \frac{2}{(4.176 \times 10^6)(0.8825)}[4.115x^3 - 0.03125x^4]\Big|_0^{15}$$

    $$+ \frac{2}{(4.176 \times 10^6)(1.3117)}[4.115x^3 - 0.03125x^4]\Big|_{15}^{35}$$

    $$= 0.00334 + 0.02140 = 0.0247 \text{ ft} = 0.30 \text{ in.}$$

    (The integration could also be done using Table 10.1, if preferred.)

Question 10.3

Since there is no sidesway involved, moment distribution is a suitable method for this problem. Since the DOF = 3, and the DSI = 3 (for vertical loading only), the stiffness and flexibility methods may also be used if preferred. This solution uses moment distribution.

1. Compute fixed-end moments.

$$\text{FEM}_{AB} = -\text{FEM}_{BA} = \text{FEM}_{CD} = -\text{FEM}_{DC} = \frac{wL^2}{12} = \frac{4(10)^2}{12} = 33.33 \text{ ft-kips}$$

$$\text{FEM}_{BC} = -\text{FEM}_{CB} = \frac{PL}{8} = \frac{40(20)}{8} = 100 \text{ ft-kips}$$

2. Compute stiffnesses and distribution factors.

$$K_{AB} = \frac{4EI}{10} = 0.4EI \qquad\qquad K_{BC} = \frac{4(2EI)}{20} = 0.4EI$$

$$K_{CD} = \frac{3EI}{10} = 0.3EI$$

$$\text{DF}_{BA} = \text{DF}_{BC} = \frac{0.4}{0.4 + 0.4} = 0.5 \qquad \text{DF}_{CB} = \frac{0.4}{0.4 + 0.3} = 0.571$$

$$\text{DF}_{CD} = 1 - 0.571 = 0.429$$

3. Perform moment distribution calculations.

Joint	A	B		C		D
End	AB	BA	BC	CB	CD	DC
DF		0.5	0.5	0.571	0.429	
FEM	33.33	−33.33	100	−100	33.33	−33.33
Bal. D						33.33
C.O.					16.67	
Bal. C				28.55	21.45	
C.O.			14.28			
Bal. B		−40.48	−40.48			
C.O.	−20.24			−20.24		
Bal. C				11.56	8.68	
C.O.			5.78			
Bal. B		−2.89	−2.89			
C.O.	−1.44			−1.44		
Bal. C				0.82	0.62	
C.O.			0.41			
Bal. C		−0.21	−0.21			
Sum	11.65	−76.91	76.89	−80.75	80.75	0.0
Check		$\Sigma = -0.02$		$\Sigma = 0.0$		

4. Sketch moment diagram.

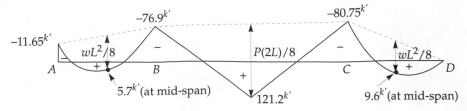

Question 10.4

Since the frame can undergo sidesway, moment distribution would be tedious. The DOF = 3 and the DSI = 3. If the flexibility method was used, the required displacements would have to be found by the unit load method, since the use of Table 10.2 is not straightforward. Therefore, the stiffness method is perhaps most suitable. The steps are similar to those in Example 10.12.

1. The DOF are the rotations at B and C, and the horizontal (rightward) displacement at B.

2. Determine the equivalent joint loads. Use units of kips and inches. $w = 2$ k/ft = 0.167 k/in.

   (a) The fixed-end moments due to loads are:

   $$\text{FEM}_{BC} = -\text{FEM}_{CB} = \frac{wL^2}{12} = \frac{0.167(180)^2}{12} = 450 \text{ in-kips}$$

   (b) Since axial deformations are neglected, the support settlement at D causes joint C to also move downwards by 0.5". The following additional fixed-end moments are therefore generated in member BC due to support movement:

   $$\text{FEM}_{BC} = \text{FEM}_{CB} = \frac{6EI\Delta}{L^2} = \frac{6(29000)(500)(0.5)}{180^2} = 1342.6 \text{ in-kips}$$

   (c) The total fixed-end moments are therefore, $\text{FEM}_{BC} = 450 + 1342.6 = 1792.6$ in-kips and $\text{FEM}_{CB} = -450 + 1342.6 = 892.6$ in-kips, and the equivalent joint loads are:

   $$\mathbf{q} = \begin{bmatrix} M_B \\ M_C \\ F_B \end{bmatrix} = \begin{bmatrix} 0 \\ 0 \\ 10 \end{bmatrix} - \begin{bmatrix} 1792.6 \\ 892.6 \\ 0 \end{bmatrix} = \begin{bmatrix} -1792.6 \text{ in-kips} \\ -892.6 \text{ in-kips} \\ 10 \text{ kips} \end{bmatrix}$$

3. Determine the stiffness matrix. Using units of kips and inches

$$\mathbf{K} = \begin{bmatrix} \frac{4EI}{L_{AB}} + \frac{4EI}{L_{BC}} & \frac{2EI}{L_{BC}} & \frac{6EI}{L_{AB}^2} \\ \frac{2EI}{L_{BC}} & \frac{4EI}{L_{BC}} + \frac{4EI}{L_{CD}} & \frac{6EI}{L_{CD}^2} \\ \frac{6EI}{L_{AB}^2} & \frac{6EI}{L_{CD}^2} & \frac{12EI}{L_{AB}^3} + \frac{12EI}{L_{CD}^3} \end{bmatrix} = EI \begin{bmatrix} 805600 & 161100 & 6042 \\ 161100 & 805600 & 6042 \\ 6042 & 6042 & 201 \end{bmatrix}$$

4. Solve for the unknown displacements.

$$\mathbf{u} = \begin{bmatrix} \theta_B \\ \theta_C \\ \Delta_B \end{bmatrix} = \mathbf{K}^{-1}\mathbf{q} = \begin{bmatrix} -3.417\times10^{-3} \text{ rad} \\ -2.020\times10^{-3} \text{ rad} \\ 0.2128 \text{ in.} \end{bmatrix}$$

5. Determine the member end-forces using Eq. 10.4.3.

   (a) For member AB the fixed-end forces are zero and therefore:

$$
\begin{bmatrix} V_A \\ M_A \\ V_B \\ M_B \end{bmatrix} = \frac{29000(500)}{120^3} \begin{bmatrix} 12 & 6(120) & -12 & 6(120) \\ 6(120) & 4(120)^2 & -6(120) & 2(120)^2 \\ -12 & -6(120) & 12 & -6(120) \\ 6(120) & 2(120)^2 & -6(120) & 4(120)^2 \end{bmatrix} \begin{bmatrix} 0 \\ 0 \\ -0.2128 \\ -3.417\times10^{-3} \end{bmatrix} = \begin{bmatrix} 0.8 \\ 460 \\ -0.8 \\ 366 \end{bmatrix}
$$

(b)   For member *BC*, the fixed-end reactions due to loads are $V_B{}^F = V_C{}^F = wL/2 = 15$ kips. The support movement effect at *C* is included in the member displacements.

$$
\begin{bmatrix} V_B \\ M_B \\ V_C \\ M_C \end{bmatrix} = \begin{bmatrix} 15 \\ 450 \\ 15 \\ -450 \end{bmatrix} + \frac{29000(500)}{120^3} \begin{bmatrix} 12 & 6(180) & -12 & 6(180) \\ 6(180) & 4(180)^2 & -6(180) & 2(180)^2 \\ -12 & -6(180) & 12 & -6(180) \\ 6(180) & 2(180)^2 & -6(180) & 4(180)^2 \end{bmatrix} \begin{bmatrix} 0 \\ -3.417\times10^{-3} \\ -0.5 \\ -2.020\times10^{-3} \end{bmatrix} = \begin{bmatrix} 15.3 \\ 366 \\ 14.7 \\ -309 \end{bmatrix}
$$

(c)   For member *AB* the fixed-end forces are zero. Rotating the member 90° clockwise:

$$
\begin{bmatrix} V_D \\ M_D \\ V_C \\ M_C \end{bmatrix} = \frac{29000(500)}{120^3} \begin{bmatrix} 12 & 6(120) & -12 & 6(120) \\ 6(120) & 4(120)^2 & -6(120) & 2(120)^2 \\ -12 & -6(120) & 12 & -6(120) \\ 6(120) & 2(120)^2 & -6(120) & 4(120)^2 \end{bmatrix} \begin{bmatrix} 0 \\ 0 \\ -0.2128 \\ -2.020\times10^{-3} \end{bmatrix} = \begin{bmatrix} 9.2 \\ 797 \\ -9.2 \\ 309 \end{bmatrix}
$$

6.   Draw the shear force and moment diagram, being careful to change from the sign convention used for the member end-forces to that used for the diagrams.

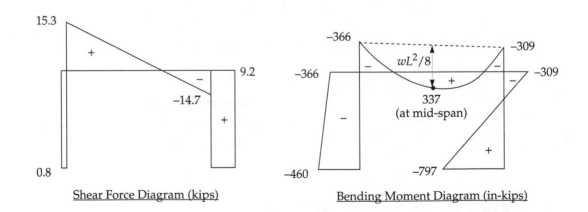

Shear Force Diagram (kips)   Bending Moment Diagram (in-kips)

Question 10.5

Moment distribution is unsuitable since the frame can undergo sidesway, and the inclined member introduces additional complications. The DOF = 3 (rotations at *B* and *C* and horizontal translation at *B*), so the stiffness method could be used, but the inclined member introduces some complications. The DSI = 1 and hence the flexibility method is perhaps the most suitable.

1. The primary structure is obtained by removing the support at *C*. The up-ward reaction at *C*, R_C, is the redundant.

2. Compute the displacement corresponding to the redundant and due to loads, and the flexibility coefficient. We use the unit load method and Table 10.1.

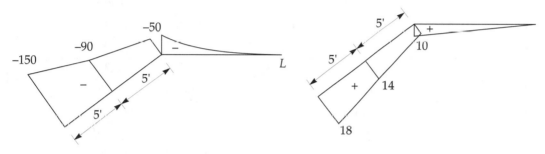

BMD due to load, M (ft-kips)   BMD due to unit value of redundant, *m*

$$u = \frac{1}{EI}\int Mm\,dx = \frac{1}{EI}\left[\frac{5}{6}[\ 150(2\times18+14)\quad 90(18+2\times14)]\right.$$

$$\left.\quad + \frac{5}{6}[\ 90(2\times14+10)\quad 50(14+2\times10)] + \frac{10}{4}(-50)(10)\right]$$

$$= -\frac{15217}{EI}$$

$$F = \frac{1}{EI}\int mm\,dx = \frac{1}{EI}\left[\frac{10}{6}[18(2\times18+10)+10(18+2\times10)]+\frac{10}{3}(10)(10)\right] = \frac{2347}{EI}$$

3. Obtain the redundant.

$$q = R_C = -\frac{u}{F} = \frac{15217}{2347} = 6.48 \text{ kips}$$

4. The reactions at *A* are obtained through equilibrium considerations. The vertical reaction is $R_A = 8.5$ kips, and the counterclockwise moment reaction is $M_A = 33$ ft-kips.

Question 10.6

As with all truss problems, we solve this using the flexibility method.

1. Cut the diagonal bars *AE* and *CE* to obtain the primary structure. Since the original structure is symmetric, by choosing redundants that preserve symmetry some work can be saved.

2. Solve the primary truss to determine all the bar forces when loads are applied and when a unit value of each redundant is applied (i.e., bar forces P_i, p_{1i}, and p_{2i}, respectively). These are tabulated below. Due to symmetry, $\Sigma p_2^2 L/A$ is identical to $\Sigma p_1^2 L/A$.

Bar	L	A	P (kips)	p_1 (kips)	p_2 (kips)	$P + F_{AE}p_1 + F_{CE}p_2$ (kips)
AB	$4L$	A	$-0.5P$	$-0.8P$	0	$-0.52P$
BC	$4L$	A	0	0	$-0.8P$	$-0.38P$
DE	$4L$	A	$0.917P$	$-0.8P$	0	$0.90P$
EF	$4L$	A	$0.917P$	0	$-0.8P$	$0.54P$
AD	$3L$	A	$-P$	$-0.6P$	0	$-1.02P$
BE	$3L$	A	0	$-0.6P$	$-0.6P$	$-0.30P$
CF	$3L$	A	0	0	$-0.6P$	$-0.28P$
BD	$5L$	$2A$	$-0.521P$	1	0	$-0.49P$
BF	$5L$	$2A$	$-1.146P$	0	1	$-0.68P$
AE	$5L$	$2A$	0	1	0	$0.026P$
CE	$5L$	$2A$	0	0	1	$0.47P$

Bar	Pp_1L/A	Pp_2L/A	p_1^2L/A	p_1p_2L/A
AB	$1.6PL/A$	0	$2.56L/A$	0
BC	0	0	0	0
DE	$-2.93PL/A$	0	$2.56L/A$	0
EF	0	$-2.93PL/A$	0	0
AD	$1.8PL/A$	0	$1.08L/A$	0
BE	0	0	$1.08L/A$	$1.08L/A$
CF	0	0	0	0
BD	$-1.3PL/A$	0	$2.5L/A$	0
BF	0	$-2.86PL/A$	0	0
AE	0	0	$2.5L/A$	0
CE	0	0	0	0
Σ	$-0.83PL/A$	$-5.79PL/A$	$12.28L/A$	$1.08L/A$

The displacements corresponding to the redundants and due to loads, **u**, and the flexibility matrix, **F**, are

$$\mathbf{u} = \frac{1}{E}\begin{bmatrix} \sum_i \dfrac{P_i p_{1i} L_i}{A_i} \\[2mm] \sum_i \dfrac{P_i p_{2i} L_i}{A_i} \end{bmatrix} = \frac{PL}{AE}\begin{bmatrix} -0.83 \\ -5.79 \end{bmatrix}$$

$$\mathbf{F} = \frac{1}{E}\begin{bmatrix} \sum_i \dfrac{p_{1i}^2 L_i}{A_i} & \sum_i \dfrac{p_{1i}p_{2i}L_i}{A_i} \\[2mm] \sum_i \dfrac{p_{1i}p_{2i}L_i}{A_i} & \sum_i \dfrac{p_{2i}^2 L_i}{A_i} \end{bmatrix} = \frac{L}{EA}\begin{bmatrix} 12.28 & 1.08 \\ 1.08 & 12.28 \end{bmatrix}$$

3. The redundants are

$$\mathbf{q} = \begin{bmatrix} F_{AE} \\ F_{CE} \end{bmatrix} = -\mathbf{F}^{-1}\mathbf{u} = \begin{bmatrix} 0.0263P \\ 0.469P \end{bmatrix}$$

4. The bar forces in the original truss can be obtained as

$$f_i = P_i + F_{AE}p_{1i} + F_{CE}p_{2i}$$

and are given in the last column of the first table.

Question 10.7

This problem is solved by the flexibility method since the DSI = 1.

1. Obtain the primary structure by removing the spring at B. The redundant force is the upward reaction at B.

2. Using Table , the displacement of the primary structure corresponding to the redundant and due to loads is

$$u = -\frac{5wL^4}{384EI} - \frac{PL^3}{48EI} = -\frac{5(1.5)/12(40\times12)^4}{384(29000)(5000)} - \frac{10(40\times12)^3}{48(29000)(5000)} = -0.755 \text{ in}$$

The flexibility coefficient due to a unit upward load at B is

$$F = \frac{1(40\times12)^3}{48(29000)(5000)} = 1.589\times10^{-2} \text{ in}$$

3. The compatibility equation must account for the deformation of the spring. If the reaction at B is R_B, then the compression in the spring is R_B/k, and equating the upward displacements we obtain

$$u + FR_B = -\frac{R_B}{k}$$

which yields

$$R_B = -\frac{u}{F + \frac{1}{k}} = \frac{0.755}{1.589 \times 10^{-2} + \frac{1}{50}} = 21.0 \text{ kips}$$

4. Using symmetry, the reactions at A and C are

$$R_A = R_C = \frac{1.5(40) + 10 - 21.0}{2} = 24.5 \text{ kips}$$

5. The vertical downward displacement at B is equal to the compression in the spring

$$\Delta_B = \frac{R_B}{k} = \frac{21}{50} = 0.42 \text{ in}$$

Question 10.8

By taking advantage of the symmetry of the structure and the loading, only one-quarter of the culvert needs to be analyzed and the problem may be reduced to that shown on the left below. This structure has a DSI of one and can be analyzed easily by the flexibility method. The structure has 3 DOF, making the stiffness method more involved. Since joints E and F can translate, moment distribution is also unsuitable. The primary structure chosen is shown on the right.

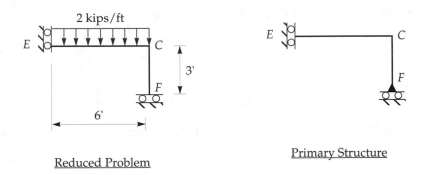

Reduced Problem

Primary Structure

1. Determine the bending moment diagrams due to external load and a unit counter-clockwise moment at F on the primary structure. These are shown below:

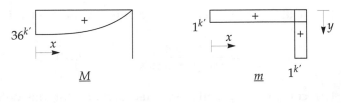

2. Compute the counter-clockwise rotation at F due to external load and the flexibility coefficient. Considering a 12" length of the culvert, the moment of inertia of the cross section is

$$I = \frac{bh^3}{12} = \frac{12 \times 8^3}{12} = 512 \text{ in}^4$$

The flexural rigidity is

$$EI = 3000 \times 512 = 1,536,000 \text{ k-in}^2 = 10,667 \text{ k-ft}^2$$

The rotation at F due to the load is

$$\theta = \frac{1}{EI}\int Mmdx = \frac{1}{EI}\int_0^6 (36-x^2)(1)dx = \frac{1}{10,667}\left[36x - \frac{x^3}{3}\right]\Bigg|_0^6 = 0.0135 \text{ rad}$$

The flexibility coefficient is

$$f = \frac{1}{EI}\int m^2 dx = \frac{1}{EI}\left(\int_0^6 dx + \int_0^3 dy\right) = \frac{9}{10,667} = 8.438\times10^{-4} \text{ rad}$$

Note: The integral could be obtained using Table 10.1.

3. Use compatibility to compute the value of the redundant force.

$$M_F = -\frac{\theta}{f} = -\frac{0.0135}{8.438\times10^{-4}} = -16 \text{ ft-kips}$$

4. Determine all other reactions using equilibrium.

$R_E = 0$ (horizontal), $M_E = 20$ ft-kips (clockwise), $R_F = 12$ kips (upward)

5. Determine the axial force, shear force and bending moment variations in BC and CD.

Axial force in BC – 0. Axial force in CD – 12 kip. Shear force in CD – 0.

The shear force diagram for BC, and the bending moment diagram for the segment BCD are shown below

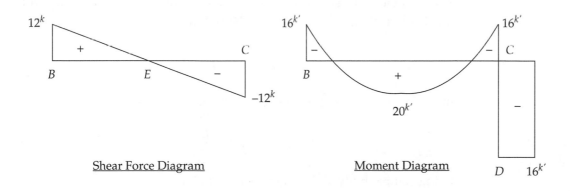

Shear Force Diagram   Moment Diagram

Question 10.9

Since the frame cannot undergo sidesway, moment distribution is suitable. Further, since joints A and C are pinned, no iteration is necessary. The DSI=1 and the DOF = 3, so the flexibility method may also be used. In the flexibility method, if the primary structure is obtained by putting a hinge at B, then each member behaves like a simple beam and Table can be used to obtain displacements due to loads. This solution outlines the use of moment distribution as well as the flexibility method.

Moment Distribution

1. Compute fixed-end moments due to loads.

$$\text{FEM}_{AB} = -\text{FEM}_{BA} = \frac{PL}{8} = \frac{wL^2}{8} \qquad \text{FEM}_{BC} = -\text{FEM}_{CB} = \frac{wL^2}{12}$$

2. Compute fixed-end moments due to support settlement.

$$\text{FEM}_{BC} = \text{FEM}_{CB} = \frac{6EI\Delta}{L^2} = \frac{6EI}{L^2}\left[\frac{wL^4}{48EI}\right] = \frac{wL^2}{8}$$

3. Unlock A and C, carry over moments to B and add with the FEM.

$$M_{BA} = -\frac{wL^2}{8} + \frac{1}{2}\left[-\frac{wL^2}{8}\right] = -\frac{3wL^2}{16}$$

$$M_{BC} = \left[\frac{wL^2}{12} + \frac{wL^2}{8}\right] + \frac{1}{2}\left[\frac{wL^2}{12} - \frac{wL^2}{8}\right] = \frac{3wL^2}{16}$$

Flexibility Method

1. The primary structure is obtained by putting a hinge at B. The redundant consists of the internal moment at B and is taken as a pair of moments, counter-clockwise at end BA and clockwise at end BC.

2. Using Table , the displacement corresponding to the redundant and due to loads is

$$u_1 = \theta_{BA} + \theta_{BC} = \frac{PL^2}{16EI} + \frac{wL^3}{24EI} = \frac{5wL^3}{48EI}$$

3. The settlement of C simply causes a rigid body rotation of member BC, and the displacement corresponding to the redundant is

$$u_2 = \theta_{BC} = \frac{wL^3}{48EI}$$

4. Imposing a unit pair of moments at B, the flexibility is

$$f = \frac{L}{3EI} + \frac{L}{3EI} = \frac{2L}{3EI}$$

5. Using compatibility, the value of the redundant is

$$q = M_B = -\frac{u_1 + u_2}{f} = -\frac{3}{16}wL^2$$

6. The bending moment diagram is

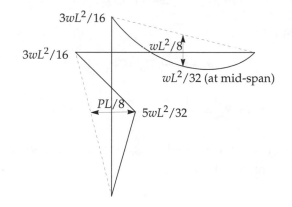

Question 10.10

As with all truss problems, we use the flexibility method.

1. The DSI = 1. The primary structure is obtained by removing the support at *B*, and the upward reaction at *B* is taken as the redundant force.

2. The truss bar forces due to the 30-kips load are denoted by P_i, and those due to a unit upward force at *B* are denoted by p_i. The bar forces and derived quantities are shown in the table below.

Bar	L (in.)	P (kips)	p (kips)	PpL	p^2L	$P + R_B p$ (kips)
AB	240	15	−0.667	−2400	106.67	15.53
AD	300	18.75	0.833	4687.5	208.33	18.09
BC	240	15	0.667	2400	106.67	15.53
CD	300	18.75	0.833	4687.5	208.33	19.41
BD	180	0	−1	0	180	0.79
Σ				−4800	810	

3. The upward displacement at *B* due to the 30-kip load is

$$u_1 = \frac{1}{EA}\sum_i P_i p_i L_i = -\frac{4800}{29000(0.5)} = -0.331 \text{ in}$$

4. A settlement at *C* of the primary structure causes it to rotate clockwise about *A* as a rigid body. The upward displacement at *B* of the primary structure due to a settlement of 0.25 in at *C* is

$$u_2 = -\frac{0.25}{2} = -0.125 \text{ in}$$

5. The flexibility coefficient is

$$f = \frac{1}{EA}\sum_i p_i^2 L = \frac{810}{29000(0.5)} = 0.055862 \text{ in}$$

6. Compatibility at B yields

$$u_1 + u_2 + fR_B = -0.5 \text{ in}$$

$$\Rightarrow R_B = \frac{-0.5 - u_1 - u_2}{f} = -0.79 \text{ kips}$$

7. The bar forces in the original truss are $P_i + R_B p_i$, and are tabulated in the last column of the table above.

Multiple Choice Problem Solutions

Questions 10.11–10.13

10.11   **d)**   Using moment distribution and Fig. 10.1, the fixed-end moments are

$$\text{FEM}_{AB} = \frac{2(15)^2}{12} + \frac{10(5)(10)^2}{15^2} + \frac{20(10)(5)^2}{15^2} = 81.94 \text{ ft-kips}$$

$$\text{FEM}_{BA} = -\frac{2(15)^2}{12} - \frac{10(5)^2(10)}{15^2} - \frac{20(10)^2(5)}{15^2} = -93.06 \text{ ft-kips}$$

Unlocking B and carrying half the moment to A yields the moment at A:

$$M_A = 81.94 + \frac{93.06}{2} = 128.5 \text{ ft-kips}$$

10.12   **c)**   Moment equilibrium about B yields:

$$128.47 - R_A(15) + 2(15)(7.5) + 10(10) + 20(5) = 0$$

$$\therefore R_A = 36.90 \text{ kips}$$

10.13   **d)**   The shear force diagram is shown below:

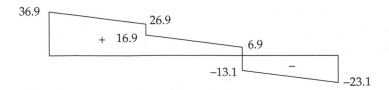

The maximum positive moment occurs at the point where the shear force diagram crosses the horizontal axis, under the 20 kip load. The moment is equal to the area under the shear force diagram from the right support to the 20 kip load:

$$\text{Maximum positive moment} = \left(\frac{13.1 + 23.1}{2}\right)(5) = 90.5 \text{ ft-kips}$$

Questions 10.14–10.18

10.14   **b)**   Use moment distribution. The fixed-end moments are:

$$\text{FEM}_{AB} = -\text{FEM}_{BA} = \frac{2(15)^2}{12} = 37.5 \text{ ft-kips}$$

$$\text{FEM}_{BC} = -\text{FEM}_{CB} = \frac{10(10)}{8} = 12.5 \text{ ft-kips}$$

The stiffnesses are: $K_{AB} = \dfrac{3EI}{15} = 0.2$, $K_{BC} = \dfrac{3EI}{10} = 0.3$

The distribution factors at B are: $\text{DF}_{BA} = \dfrac{0.2}{0.2 + 0.3} = 0.4$,

$\text{DF}_{BC} = 1 - 0.4 = 0.6$

Perform moment distribution calculations.

Joint	A	B		C
End	AB	BA	BC	CB
DF		0.4	0.6	
FEM	37.5	−37.5	12.5	−12.5
Bal. A	−37.5			
C.O.		18.75		
Bal. C				12.5
C.O.			6.25	
Bal. B		15	22.5	
Σ	0	−41.25	41.25	0

10.15   **a)**   Taking moments about B for the free body AB yields

$$R_A(15) - 2(15)(7.5) + 41.25 = 0$$

$$\therefore R_A = 12.25 \text{ kips}$$

10.16   **c)**   Taking moments about B for the free body BC yields

$$R_C(10) - 10(5) + 41.25 = 0$$

$$\therefore R_C = 0.875 \text{ kips}$$

Vertical equilibrium of the entire structure yields $R_B = 2(15) + 10 - R_A - R_C = 26.9$ kips.

10.17   **a)**   The shear in span AB is zero at

$$x = \frac{R_A}{w} = \frac{12.25}{2} = 6.125 \text{ ft}$$

The maximum moment in AB occurs at this point and is

$$M_{max} = R_A x - w\frac{x^2}{2} = 12.25(6.125) - 2\frac{6.125^2}{2} = 37.52 \text{ ft-kips}$$

10.18   **a)**   The maximum positive moment occurs at the 10 kip load application point.

$$M_{max} = R_C(5) = 0.875(5) = 4.375 \text{ ft-kips}$$

Questions 10.19–10.24

10.19   **c)**   Use moment distribution. The fixed-end moments are:

$$\text{FEM}_{AB} = -\text{FEM}_{BA} = \frac{15(10)}{8} = 18.75 \text{ ft-kips}$$

$$\text{FEM}_{BC} = -\text{FEM}_{CB} = \frac{1(10)^2}{12} = 8.33 \text{ ft-kips}$$

Stiffnesses and DF's are: $K_{AB} = \dfrac{4EI}{L}$, $K_{BC} = \dfrac{3EI}{L}$, $\text{DF}_{BA} = \dfrac{4}{7}$, $\text{DF}_{BC} = \dfrac{3}{7}$.

Perform moment distribution calculations.

Joint	A	B		C
End	AB	BA	BC	CB
DF		4/7	3/7	
FEM	18.75	−18.75	8.33	−8.33
Bal. C				8.33
C.O.			4.17	
Bal. B		3.57	2.68	
C.O.	1.79			
Σ	20.54	−15.18	15.18	0

10.20   **d)**

10.21   **a)**   Taking moments about B of member AB yields:

$$R_{A_x}(10) - 15(5) + 15.18 = 0$$

$$\therefore R_{A_x} = 5.98 \text{ kips}$$

10.22   **c)**   The upward reaction at A is equal to the shear at B in member BC. Taking moments about C in member BC yields:

$$V_B(10) - 15.18 - 1(10)(5) = 0$$

$$\therefore V_B = R_{A_Y} = 6.518 \text{ kips}$$

10.23   **b)**   The upward reaction at C: $R_C = 10 - 6.518 = 3.482$ kips.
The shear in BC is zero at a distance of 3.482 ft from C. The moment at this point is

$$M_{max} = R_C(3.482) - 1\frac{(3.482)^2}{2} = 6.06 \text{ ft-kips}$$

10.24   **e)** $M = R_{A_X}(5) = 5.98(5) = 29.9$ ft-kips

Water Treatment

by Mackenzie L. Davis

Chapter 11

Solutions to Practice Problems

11.1   Calculate the mmoles of alum ($Al_2(SO_4)_3 \cdot 14H_2O$):

$$\frac{40 \text{ mg/L}}{594 \times 10^3 \text{ mg/mole}} = 6.73 \times 10^{-5} \text{ moles/L}$$

This is equivalent to 0.0673 millimoles/L. From the reaction given in Eq. 11.3.3, calculate the mmoles of alkalinity consumed. Since 6 moles of HCO_3 are consumed for each mole of alum:

$$6(0.0673) = 0.4040 \text{ mmoles/L of alkalinity}$$

Convert to mg/L:

(0.4040 mmoles/L)(61 mg/mmole) = 24.6 mg/L as HCO_3 are consumed

Convert the initial alkalinity to mg/L as HCO_3. The equivalent weight of $CaCO_3$ is

$$\frac{GMW}{n} = \frac{100.09}{2} = 50.04$$

The equivalent weight of HCO_3 is

$$\frac{61.02}{1} = 61.02$$

The mg/L of HCO_3 is

Essay Problems

$$\left(16.39 \text{ mg / L as } CaCO_3\right)\left(\frac{\text{EW of Species}}{\text{EW of } CaCO_3}\right)$$

$$(16.39)\left(\frac{61.02}{50.04}\right) = 19.98 \text{ or } 20 \text{ mg / L}$$

Since 20 mg/L is less than the 24.6 mg/L required, there is not enough alkalinity.

11.2   Compute the volume of the basin:

$$V = (30.75)(24.75)(61.5) = 46,805 \text{ ft}^3$$

In SI units:

$$V = (46,805 \text{ ft}^3)(0.02832 \text{ m}^3/\text{ft}^3) = 1326 \text{ m}^3$$

The viscosity of water at 10 °C is 1.307×10^{-3} Pa·s. The power input required is

$$P = (1.307 \times 10^{-3})(35)^2(1326) = 2123 \text{ or } 2120 \text{ W}$$

For two propellers on separate shafts, the power for one shaft is:

$$0.5(2120) = 1060 \text{ W}$$

Select a $K = 1.00$ from Table 11.4. Select a D using the assumption that it is 30% of the width:

$$D = (0.30)(30.75 \text{ ft})(0.3048 \text{ m/ft}) = 2.81 \text{ m}$$

Solve for n in Eq. 11.3.9:

$$n = \left[\frac{P}{K\rho D^5}\right]^{1/3} = \left[\frac{1,060}{(1.00)(1000)(2.81)^5}\right]^{1/3} = 0.182 \text{ rps or } 10.9 \text{ rpm}$$

11.3   Convert mg/L as the ion to mg/L as $CaCO_3$.

Constituent	mg/L	EW	EW $CaCO_3$/EW	mg/L as $CaCO_3$	mEq/L
CO_2	42.7	22.0	2.28	97.36	1.94
Ca^{2+}	102.0	20.0	2.50	255	5.10
Mg^{2+}	45.2	12.2	4.12	186.22	3.70
Na^+	21.8	23.0	2.18		
HCO_3	420.0	61.0	0.820	344	6.88
Cl	32.0	35.5	1.41		
SO_4	65.0	48.0	1.04		

Note that Na^+, Cl and SO_4 were not converted since they do not enter into the computations.

Calculate the split:

$$x = \frac{40-10}{186.22-10} = 0.17$$

In the first stage the water is softened to the theoretical solubility limits; lime and soda must be added as shown below.

Addition = to:	Lime mg/L as $CaCO_3$	Lime mEq/L	Soda mg/L as $CaCO_3$	Soda mEq/L
CO_2	97.36	1.94		
HCO_3	344	6.88		
Mg	186.22	3.70		
(Ca + Mg)– HCO_3			97.22	1.92
	627.58	12.52	97.22	1.92

The fraction of water passing through the first stage is $1 - 0.17 = 0.83$. The total hardness of the water after passing through the first stage is the theoretical solubility limit, i.e., 40 mg/L as $CaCO_3$. Since the total hardness in the raw water is $255 + 186 = 441$ mg/L as $CaCO_3$, the mixture of the treated and bypass water has a hardness of

$$0.17(441) + 0.83(40) = 108 \text{ mg/L as } CaCO_3$$

This is within the acceptable range of 80 to 120 mg/L as $CaCO_3$, so no further treatment is required.

The split treatment flow scheme is shown below.

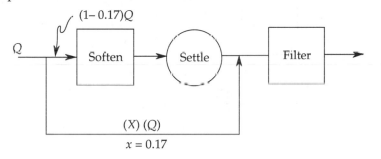

11.4 In the first stage the water is softened to the theoretical solubility limits; lime and soda must be added as shown below.

Addition = to:	Lime mEq/L	Soda mEq/L
CO_2	0.40	
HCO_3	2.72	
Mg	1.12	
(Ca + Mg)– HCO_3		0.56
	4.24	0.56

Convert mEq/L of Mg to mg/L as $CaCO_3$:

$$(1.12 \text{ mEq/L})(50 \text{ mg } CaCO_3/\text{mEq}) = 56 \text{ mg/L as } CaCO_3$$

The split is

$$x = \frac{40 - 10}{56 - 10} = 0.65$$

The fraction of water passing through the first stage is $1 - 0.65 = 0.35$. The total hardness of the water after passing through the first stage is the theoretical solubility limit, 40 mg/L as $CaCO_3$. Multiply the equivalent weight of $CaCO_3$ times the mEq/L to find total hardness in the raw water:

$$(2.16)(50) + (1.12)(50) = 164 \text{ mg/L as } CaCO_3$$

The mixture of the treated and bypass water has a hardness of

$$0.65(164) + .35(40) = 120.6 \text{ mg/L as } CaCO_3$$

This is within the acceptable range of 80 to 120 mg/L as $CaCO_3$, so no further treatment is required.

The chemical dose in mg/L as CaO and Na_2CO_3 is

Lime $= (4.24 \text{ mEq/L})(28 \text{ mg CaO/mEq}) = 118.72$ or 120 mg/L as CaO

Soda $= (0.56 \text{ mEq/L})(53 \text{ mg } Na_2CO_3/\text{mEq}) = 29.7$ or 30 mg/L as Na_2CO_3

The split treatment flow scheme is shown below:

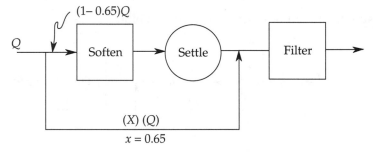

11.5   The total surface area of the tanks is

$$\frac{\left(0.800 \text{ m}^3/\text{s}\right)\left(86\,400 \text{ s/d}\right)}{15 \text{ m}^3/\text{d} \cdot \text{m}^2} = 4608 \text{ m}^2$$

With a L/W of 5:1, a single tank would be

$$(5x)(x) = 4608$$

$$x = (921.60)^{1/2} = 30.36 \text{ m wide}$$

and $5(30.36) = 151.8$ m long. This exceeds the maximum length of 100 m. If four tanks, each 15 m $\times$ 75 m were used, the total surface area would be

$$4(15)(75) = 4500 \text{ m}^2$$

The overflow rate would be

$$\frac{\left(0.800 \text{ m}^3/\text{s}\right)\left(86\,400 \text{ s/d}\right)}{4500 \text{ m}^2} = 15.36 \text{ m}^3/\text{d} \cdot \text{m}^2$$

This is sufficiently close to the desired 15 m/d.

With a side water depth of 4.5 m, the volume of the four tanks is
$$(4.5)(4500) = 20\,250 \text{ m}^3$$

The detention time is

$$t_o = \frac{20\,250 \text{ m}^3}{\left(0.800 \text{ m}^3/\text{s}\right)\left(3600 \text{ s/h}\right)} = 7 \text{ h}$$

The length of weir in each tank is

$$\frac{\left(0.800 \text{ m}^3/\text{s}\right)\left(86\,400 \text{ s/d}\right)}{(4 \text{ tanks})(150 \text{ m/d})} = 115 \text{ m}$$

To provide for the maximum day flow, provide 1.5(4) = 6 tanks.

11.6   At 4.2 hours the overflow rate is

$$v_o = \left(\frac{5.0 \text{ m}}{4.2 \text{ h}}\right)(24 \text{ h/d}) = 28.57 \text{ or } 29 \text{ m/d}$$

Begin the calculation for the percent removal by plotting a vertical line from the intersection of the 70% line and the bottom of the test column (i.e., at 4.2 h and 5.0 m) and determining the mid-point between each of the isoconcentration lines. The percent removal is then calculated as

$$R_T = 70 + \frac{3.9}{5.0}(85-70) + \frac{2.5}{5.0}(90-85) + \frac{1.1}{5.0}(100-90)$$

$$= 70 + 11.7 + 2.5 + 2.2 = 86.4\%$$

11.7   The mass of sludge generated each day is
$$M_s = (86.40)(0.10)(135.0 + (0.58)(29.9) + 237.5) = 3368 \text{ kg/d}$$

The specific gravity of the sludge is

$$S_{sl} = \frac{2.75}{0.10 + (2.75)(0.90)} = 1.07$$

The volume of the sludge is

$$V_{sl} = \frac{3368}{(1000)(1.07)(0.10)} = 31.48 \text{ m}^3/\text{d}$$

The approximate volume after dewatering is

$$V_2 = 31.48\left(\frac{0.10}{0.60}\right) = 5.25 \text{ or } 5.3 \text{ m}^3/\text{d}$$

On an annual basis the volume is
$$V = (5.25 \text{ m}^3/\text{d})(365 \text{ d/y}) = 1916 \text{ or } 1900 \text{ m}^3/\text{y}$$

11.8 A) The Head Loss Through a Clean Bed:

To compute the head loss, the last row in the table must be completed. Note that English units are used.

The Reynolds number is required to calculate the drag coefficient. Using the temperature of the water, the kinematic viscosity is found to be 1.408×10^{-5} ft^2/s. The velocity is found from the filtration rate:

$$v_a = \left(2.5 \text{ gal/min} \cdot \text{ft}^2\right)\left(0.1337 \text{ ft}^3/\text{gal}\right)\left(\frac{1}{60 \text{ s/min}}\right) = 0.00557 \text{ ft/s}$$

For a sand particle with a sphericity of 0.95 and a diameter of 0.000583 ft, the Reynolds number is

$$R = \frac{(0.95)(0.000583)(0.00557)}{1.408 \times 10^{-5}} = 0.219$$

Because the Reynolds number is less than 0.5, calculate the drag coefficient as

$$C_D = \frac{24}{0.219} = 109.6$$

The product $C_D(f)/d$ for the last row is

$$\frac{(109.6)(0.0166)}{0.000583} = 3120$$

The sum of the last column is 45 640. The head loss is

$$h_L = \frac{(1.067)(2.0)(0.00557)^2}{(0.95)(32.2)(0.40)^4}(45\,640) = 3.86 \text{ ft}$$

This head loss exceeds 0.6 m (2 ft) and would be considered excessive.

B) Height of Backwash Troughs:

To compute the depth of the expanded bed, the last row in the table must be completed. Note that English units are used.

Estimate the settling velocity of the sand particle as 1.4 cm/s using Fig. 11.9 with a diameter of 0.178 cm and a specific gravity of 2.5.

The Reynolds number is calculated as

$$R = \frac{(0.95)(0.000583)(1.4 \text{ cm})\left(\frac{1}{30.5 \text{ cm/ft}}\right)}{1.408 \times 10^{-5}} = 1.806$$

Because the Reynolds number is in the transition region, the drag coefficient is estimated as

$$C_D = \frac{24}{1.806} + \frac{3}{\sqrt{1.806}} + 0.34 = 15.86$$

The settling velocity is calculated as

$$v_s = \left[\frac{4(32.2)(2.65-1)(0.000583)}{3(15.86)}\right]^{1/2} = 0.0510 \text{ ft/s}$$

The velocity of the backwash is:

$$v_b = \left(20 \text{ gpm/ft}^2\right)\left(0.1337 \text{ ft}^3/\text{gal}\right)\left(\frac{1}{60 \text{ s/min}}\right) = 0.0446 \text{ ft/s}$$

Since v_b is less than settling velocity of the smallest sand grains, filter media will not be lost and the backwash rate is acceptable.

The porosity of the expanded bed is

$$\varepsilon_e = \left(\frac{0.0446}{0.0510}\right)^{0.22} = 0.970$$

The fraction $(f/(1-\varepsilon_e))$ for the last row is

$$\frac{0.0166}{1-0.970} = 0.5533$$

Summing the last column, the depth of the expanded bed is

$$D_e = (1-0.40)(2.0)(2.45) = 2.94 \text{ ft}$$

The bottom of the backwash trough must be

$$2.94 - 2 + 0.5 = 1.44 \text{ ft}$$

or about 1.5 ft above the surface of the sand bed.

C) Smallest Size Anthracite Coal Particle:

Using Eq. 11.6.1 and the specific gravities of the coal and sand, the diameter of the smallest coal particle must be such that it settles on top of the smallest sand particle (0.000583 ft):

$$d_A = (0.000583)\left[\frac{2.65-1.00}{1.55-1.00}\right]^{2/3} = 0.001213 \text{ ft}$$

11.9   From Table 11.10 at a temperature of 10 °C and TDS of 80 mg/L find

$$(pK_2 - pK_s) = 2.40$$

Convert Ca to moles/L:

$$\left(25.00 \text{ mg/L as CaCO}_3\right)\left(\frac{20}{50}\right) = 10.00 \text{ mg/L as Ca}$$

$$\left(10.00 \text{ mg/L}\right)\left(\frac{1}{\left(40\times10^3 \text{ mg/mole}\right)}\right) = 2.50\times10^{-4} \text{ moles/L}$$

Calculate pCa:

$$pCa = -\log(2.5 \times 10^{-4}) = 3.60$$

Convert alkalinity to moles/L assuming all alkalinity is CO_3 since the pH is > 8.3:

$$\left(40 \text{ mg} / \text{L as } CaCO_3\right)\left(\frac{30}{50}\right) = 24.0 \text{ mg} / \text{L as } CaCO_3$$

$$\left(24.0 \text{ mg} / \text{L}\right)\left(\frac{1}{60 \times 10^3 \text{ mg} / \text{mole}}\right) = 4.00 \times 10^{-4} \text{ moles} / \text{L}$$

Calculate *pAlk*:
$$pAlk = -\log(4.00 \times 10^{-4}) = 3.40$$

The stability index is
$$SI = 8.60 - 2.40 - 3.60 - 3.40 = -0.80$$

Therefore the water is corrosive. Add lime.

11.10   The available chlorine from chlorine gas is 100%. The cost of available chlorine is then $300 per ton.

For chlorine dioxide, using the half reaction from Table 11.11, the equivalent weight is

$$EW = \frac{67.5}{1} = 67.5$$

The percent available chlorine is

$$\% \text{ Available Chlorine} = \frac{35.5}{67.5}(100\%) = 52.6\%$$

The cost of available chlorine is then

$$\frac{\$600/\text{ton}}{0.526} = \$1140/\text{ton of available chlorine}$$

This is obviously much more expensive.

Other considerations are the fact that chlorine dioxide does not form THMs and that its reaction byproducts include chlorite and chlorate. While the elimination of THMs is desirable to reduce cancer risk, chlorite and chlorate compounds may have a human health risk. Chlorine dioxide also has the potential of generating taste and odor problems.

Multiple Choice Problems

11.11   **a)**   Using the national per capita water consumption of 628 Lpcd, the estimated population is

$$\frac{\left(2200 \times 10^6 \text{ gal/d}\right)\left(3.785 \text{ L/gal}\right)}{628 \text{ L/capita} \cdot \text{d}} = 13\ 259\ 554 \text{ people}$$

11.12   **d)**   The Shultz-Hardy rule states that monovalent, divalent and trivalent species should be effective approximately in the ratio $1:10^{-2}:10^{-3}$. In most practical systems, the Shultz-Hardy rule is "violated" because the electrolytes are not indifferent. Inorganic compounds added to water react with solution electrolytes and form complexes and precipitates. Hence, coagulation power for monovalent, divalent and trivalent species is taken as 1:60:700. The ratio for alum or ferric chloride would be

$$\frac{0.10 \text{ mole} / \text{L}}{700} = 1.4 \times 10^{-4} \text{ mole} / \text{L}$$

11.13 b) Significant removal of $CaCO_3$ is achieved when the pH is above about 9.6 to 10.8. Significant removal of $Mg(OH)_2$ is achieved when the pH is above about 10.8 to 11.5. Because the water quality (hardness) from the source may vary with time, controlling neither the flow rate nor the feed rate precisely guarantees that the desired pH will be achieved. Only pH control will ensure the softening precipitation process operates correctly. There is no such thing as a hardness meter.

11.14 d) Convert each of the units to mg/L as CO_2:

$$(10^{-4}\,mole/L)(44 \times 10^3\,mg/mole) = 4.40\ mg/L$$
$$(20\ mg/L\ as\ CaCO_3)(22/50) = 8.80\ mg/L$$
$$(0.40\ mEq/L)(22\ mg/mEq) = 8.80\ mg/L$$

Since 10 mg/L is the guideline for air stripping, none of these waters is a candidate for air stripping.

11.15 c) Increasing the safety factor for the sedimentation tank will not improve performance if the water cannot be coagulated/flocculated. Likewise, minimum G values will have little effect because the particles are too dispersed and are already having trouble coming into contact. Additions of polymer and coagulant aid may improve performance but sludge production will increase and cost will be high. The most appropriate solution is direct filtration. This saves the cost of building coagulation/flocculation tanks and the cost of chemicals.

11.16 b) Calculate the settling velocity in m/d:

$$(0.0579\ cm/s)\left(\frac{1}{100\ cm/m}\right)(86\,400\ s/d) = 50.03\ m/d$$

Calculate the percent removal using Eq. 11.5.10:

$$\frac{50.03}{77.5}(100\%) = 64.55 \quad or\ 64.6\%$$

11.17 b) Percent reduction may be calculated a

$$\left(\frac{v_1 - v_2}{v_1}\right)(100\%) = \left(1 - \frac{v_2}{v_1}\right)(100\%)$$

The ratio of volumes based on Eq. 11.5.15 is

$$\frac{v_2}{v_1} = \frac{P_1}{P_2} = \frac{0.10}{0.75} = 0.1333$$

The percent reduction is:

$$\%\ Reduction = (1 - 0.1333)(100\%) = 86.67\%$$

11.18 d) The saturation index is calculated as

$$SI = pH - pH_s = 10.2 - 9.80 = +0.40$$

This water will precipitate $CaCO_3$ but lime will only raise the pH further and continue to precipitate $CaCO_3$. The pH of the water must be lowered to stop the precipitation. Recarbonation with carbon dioxide will lower the pH.

11.19 **c)** Free chlorine is defined as HOCl and OCl. Only dichloramine ($NHCl_2$) will not provide free chlorine.

11.20 **a)** Because chlorine gas is more dense than air the duct inlets should be placed near the floor.

Wastewater Treatment

by Mackenzie L. Davis

Chapter 12

Solutions to Practice Problems

12.1 Note that capital K implies that the rate constants are for base 10. This problem may be worked in base 10 or the rate constants may be converted to base e. Base 10 is used in this solution.

Essay Problems

Because the deficit after mixing is 0.0, $D_a = 0$.

Calculate the travel time downstream from the discharge point:

$$t = \frac{(33.16 \text{ km})(1000 \text{ m/km})}{(0.5 \text{ m/s})(86\,400 \text{ s/d})} = 0.7676 \text{ d}$$

Convert the ultimate BOD from a mass discharge to mg/L:

$$L_a = \frac{(2324 \text{ kg/d})(1 \times 10^6 \text{ mg/kg})}{(6.500 \text{ m}^3/\text{s})(86\,400 \text{ s/d})(10^3 \text{ L/m}^3)} = 4.138 \text{ mg/L}$$

Calculate the deficit:

$$D = \frac{(1.370)(4.138)}{4.830 - 1.370}\left[10^{-(1.370)(0.7676)} - 10^{-(4.830)(0.7676)}\right] = 0.145 \text{ mg/L}$$

Calculate the DO by first determining the saturated DO from Table 12.2 using a temperature of 24° C: $DO_s = 8.53$ mg/L:

$$DO = DO_s - D$$
$$DO = 8.53 - 0.145 = 8.385 \text{ or } 8.39 \text{ mg/L}$$

12.2   From the figure, one may observe that the average flow is 2.0 MGD. The storage required may be recognized as the area under the curve from 6 am to 6 pm. The volume of the equalization basin may be calculated by numerical integration or, because of the geometric symmetry, the volume may be calculated as the area of a triangle:

$$V = \left(\frac{1}{2}\right)(b)(h) = (0.5)(12 \text{ h})\left(1 \times 10^6 \text{ gal/d}\right)\left(\frac{1 \text{ d}}{24 \text{ d}}\right) = 250,000 \text{ gal}$$

Add 25% for safety:
$$V = (1.25)(250,000) = 312,500 \text{ gal}$$

In cubic feet:
$$V = (312,500)(0.1337 \text{ ft}^3/\text{gal}) = 41,781 \text{ ft}^3$$

For a depth of 10 ft the area is:
$$A_s = \frac{41,781 \text{ ft}^3}{10 \text{ ft}} = 4,178 \text{ ft}$$

If the length and width are equal:
$$L = W = (4,178)^{1/2} = 64.63 \text{ or } 65 \text{ ft per side}$$

The minimum power required is:
$$(0.02 \text{ hp}/1000 \text{ gal})(312,500 \text{ gal}) = 6.25 \text{ hp}$$

12.3   Convert the particle density to units of kg/m^3:

$$\rho_s = \left(\frac{1.83 \text{ g}}{\text{cm}^3}\right)\left(\frac{10^{-3} \text{ kg}}{\text{g}}\right)\left(\frac{10^6 \text{ cm}^3}{\text{m}^3}\right) = 1830 \text{ kg/m}^3$$

Calculate the settling velocity using Stokes Law. The viscosity is determined using the temperature of 12° C in Table 4.2:

$$v_s = \frac{(9.80)((1830 - 1000)\left(1.71 \times 10^{-4}\right)^2}{(18)\left(1.235 \times 10^{-3}\right)} = 1.07 \times 10^{-2} \text{ m/s}$$

The time for a particle to fall through the depth of the grit chamber is:

$$t = \frac{h}{v_s} = \frac{1.0 \text{ m}}{1.07 \times 10^{-2} \text{ m/s}} = 93.5 \text{ s}$$

Since the detention time in the chamber is only 60 s, the particle will not be captured in the chamber.

12.4   Begin by calculating the new overflow rate. The surface area of the tank is computed in Example 12.4 as 170.42 m².

The new overflow rate is

$$v_o = \frac{Q}{A_s} = \frac{0.0789 + 0.0395}{170.42 \text{ m}^2}(86,400 \text{ s/d}) = 60.0 \text{ m}^3/\text{d} \cdot \text{m}^2$$

This is equivalent to 1,473 gal/d·ft². From Fig. 12.1, the percent BOD_5 removal is 27%.

Because the industrial discharge BOD is all soluble, the increased flow rate will decrease the BOD removal of the fraction associated with the municipal flow. Furthermore, the concentration of BOD in the municipal flow will be diluted since these graphs assume BOD removal is a result of suspended solids. The new equivalent BOD and suspended solids of the raw waste water is:

$$BOD = \frac{(0.0395)(0) + (135)(0.0789)}{0.0395 + 0.0789} = 90 \text{ mg/L}$$

$$SS = \frac{(0.0395)(0) + (200)(0.0789)}{0.0395 + 0.0789} = 133 \text{ mg/L}$$

Using the 27% and 90 mg/L BOD_5 from Fig. 12.2:

$$t = 1.67 \text{ h}$$

This yields 48% efficiency of suspended solids removal.

The BOD leaving the sedimentation tank is equivalent to all of the BOD from the industrial process plus the fraction of municipal BOD not removed in the primary tank, i.e.:

$$(1 - 0.27)(90 \text{ mg/L}) = 65.7 \text{ mg/L}$$

The BOD concentration is the weighted average:

$$BOD = \frac{(0.0395)(330) + (0.0789)(65.7)}{0.1184} = 153.9 \text{ or } 154 \text{ mg/L}$$

12.5   Calculate the required removal efficiency:

$$E = \frac{125 - 25}{125}(100\%) = 80\%$$

Calculate F:

$$F = \frac{1 + 12}{\left(1 + (0.1)(12)\right)^2} = 2.69$$

Solve Eq. 12.4.17 for the volume:

$$V = \frac{w}{(F)\left(\dfrac{\dfrac{100}{E_1} - 1}{0.0085}\right)^2}$$

Calculate the volume of the filter:

$$V = \frac{(125 \text{ mg/L})(1 \text{ MGD})(8.34 \text{ lb/gal})}{2.69\left[\dfrac{\dfrac{100}{80} - 1}{0.0085}\right]^2} = 0.448 \text{ acre-ft}$$

Calculate the area of the filter:

$$A = \frac{0.448 \text{ acre-ft}}{5.52 \text{ ft}} = 0.0812 \text{ acre or } 3,535 \text{ ft}^2$$

The diameter of the filter is then

$$\frac{(\pi)(D)^2}{4} = 3,535$$

$$D = 67 \text{ ft}$$

Note: The recycle ratio of 12 far exceeds the NRC optimum of 8.

12.6   The BOD_5 after primary settling is

$$S_o = (1 - 0.33)(135) = 90.45 \text{ mg/L}$$

S_i is then calculated:

$$S_i = \frac{90.45 + (0.25)(12)}{1 + 0.25} = 74.76$$

Calculate the mean cell residence time:

$$\frac{1}{\theta_c} = \frac{(2.5)(90.45 - 12)}{(90.45 - 12) + \left[(1.25)(100)\ln\left(\frac{74.76}{12}\right)\right]} - 0.05$$

$$\theta_c = 1.699 \text{ d}$$

With the MLVSS, solve for the hydraulic detention time:

$$2,500 = \frac{(1.699)(0.5)(90.45 - 12)}{\theta(1 + (0.05)(1.699))}$$

$$\theta = 0.0246 \text{ d}$$

The volume of the plug flow reactor is then

$$V = \theta Q = (0.0246 \text{ d})(0.1065 \text{ m}^3/\text{s})(86\,400 \text{ s/d}) = 226.4 \text{ or } 226 \text{ m}^3$$

12.7   Assuming 1/3 of BOD is removed in the primary tank, the pounds of BOD entering the aeration tank are

$$(2/3)(150 \text{ mg/L})(10 \text{ MGD})(8.34 \text{ lb/gal}) = 8340 \text{ lb } BOD_5/\text{d}$$

For a loading of 40 lb/1,000 ft³, the tank volume is

$$V = \frac{8,340 \text{ lb BOD/d}}{0.04 \text{ lb/ft}^3} = 208,500 \text{ ft}^3 \text{ or } 1.56 \times 10^6 \text{ gal}$$

The MLVSS is calculated from the F/M ratio and the aeration tank volume of 1.56 MG:

$$\frac{F}{M} = 0.21 = \frac{8,340}{M}$$

$$M = \frac{8,340}{0.21} = 39,714$$

$$M = (x \text{ mg/L})(1.56 \text{ MG})(8.34 \text{ lb/gal}) = 39,714$$

$$x = 3,052 \text{ or } 3,100 \text{ mg/L}$$

Calculate the air required by first determining the BOD_5 to be removed:

lb of BOD_5 to be removed = 0.85(8,340) = 7,089

Assume 800 ft³ air/lb BOD_5 removed to determine the volume of air at 0° C and 760 mm Hg:

$$(800 \text{ ft}^3/\text{lb})(7,089) = 5.67 \times 10^6 \text{ ft}^3$$

Calculate the volume of air at 20° C:

$$5.67 \times 10^6 \left(\frac{293 \text{ °C}}{273 \text{ °C}} \right) = 6.09 \times 10^6 \text{ ft}^3$$

Calculate the volume of settled sludge:

$$V_{sl} = \frac{3{,}052 \text{ mg/L}}{8{,}000 \text{ mg/L}} (1{,}000 \text{ mL}) = 381 \text{ mL or } 38.1\%$$

The sludge volume index is calculated as

$$SVI = \frac{38.1\%}{0.3052\%} = 125$$

12.8 Determine the observed yield coefficient:

$$Y_{obs} = \frac{0.5}{1 + (0.05)(8)} = 0.3571$$

Determine the increase in mass of MLVSS:

$$P_x = (0.3571)(0.250 \text{ m}^3/\text{s})(250 \text{ g/m}^3 - 6 \text{ g/m}^3)(86\,400 \text{ s/d})(10^{-3} \text{ kg/g}) = 1882 \text{ kg/d}$$

The theoretical oxygen requirement is

$$\frac{(0.250)(250 - 6)(86\,400)(10^{-3})}{0.68} - (1.42)(1882) = 5078 \text{ kg/d}$$

Assuming air has a density of 1.185 kg/m^3 and contains 23.2% oxygen by mass, the theoretical air required is

$$\frac{5078}{(1.185)(0.232)} = 18\,470 \text{ m}^3/\text{d}$$

With an oxygen transfer efficiency of 8% and a safety factor of 2:

$$\frac{18\,471}{0.08}(2) = 461\,800 \text{ m}^3/\text{d}$$

In units of m^3/s:

$$\frac{461\,800}{86\,400} = 5.34 \text{ m}^3/\text{s}$$

12.9 Calculate the solids removed in the primary settling:

(1 MG)(8.34 lb/gal)(0.53)(200 mg/L) = 884 lb/d dry solids

Assuming the primary sludge water content = 95%, the lbs of water sent to the digester is

$$\frac{884}{0.05} = \frac{x}{0.95}$$

$$x = 16{,}800 \text{ lb water/d}$$

The volume of sludge pumped to the digester is then

$$\frac{884 \text{ lbs}}{(1.10)(62.4 \text{ lb/ft}^3)} + \frac{16{,}800 \text{ lbs}}{62.4 \text{ lb/ft}^3} = 282 \text{ ft}^3/\text{d}$$

The solids removed in the secondary settling tank are

$$\left[(200 \text{ mg/L})(1.00-0.53)-20 \text{ mg/L}\right](1 \text{ MGD})(8.34 \text{ lb/gal}) = 617 \text{ lb}$$

where $(1.00 - 0.53)$ = the suspended solids removed in the primary settling tank and the 20 mg/L is the effluent suspended solids from the secondary clarifier.

Assuming the secondary sludge water content = 98%, the lbs of water sent to the digester is:

$$\frac{617}{0.02} = \frac{x}{0.98}$$

$$x = 30,230 \text{ lb}$$

The volume pumped to the digester is then

$$\frac{617}{(1.10)(62.4)} + \frac{30,230}{62.4} = 493 \text{ ft}^3/\text{d}$$

The wastage from the activated sludge process is

$$(0.50)(1.00-0.333)(300 \text{ mg/L})(1 \text{ MGD})(8.34 \text{ lb/gal}) = 834 \text{ lb/d}$$

where (0.5) is the fraction of BOD synthesized and $(1.00 - 0.33)$ = the fraction of BOD removed in the primary tank.

With a water content of 98%, the lbs of water pumped to the digester is

$$\frac{834}{0.02} = \frac{x}{0.98}$$

$$x = 40,900 \text{ lb/d}$$

The volume of sludge pumped is then

$$\frac{834}{(1.10)(62.4)} + \frac{40,900}{62.4} = 667 \text{ ft}^3$$

The total volume of sludge pumped to the digester is then

$$V_{sl} = 282 + 493 + 667 = 1,442 \text{ ft}^3/\text{d}$$

With a 15-day detention time, the storage volume is

$$(1,442 \text{ ft}^3/\text{d})(15 \text{ d}) = 21,630 \text{ ft}^3$$

If sludge is withdrawn once per week, an additional storage volume is required:

$$(1,442)(7 \text{ d}) = 10,090 \text{ ft}^3$$

The total sludge storage volume for the digester is then

$$21,630 + 10,090 = 31,800 \text{ ft}^3$$

The volume of digester gas is a function of the destruction of COD. Assuming that COD destroyed = BOD and that 6.35 ft³ of methane is produced for each pound of COD destroyed, we may estimate the gas production as follows:

From the primary tank the digester receives 885 lb/d of dry solids of which

$$(1/3)(300 \text{ mg/L})(1 \text{ MGD})(8.34 \text{ lb/gal}) = 834 \text{ lb/d of BOD}$$

Using the assumption that 1 lb of MLVSS = 0.6 lb of BOD, the waste activated sludge yields:

$$(834 \text{ lb/d})(0.6) = 500 \text{ lb/d of BOD}$$

The volume of methane produced is

$$(834 + 500)(6.35 \text{ ft}^3 \text{ CH}_4/\text{lb COD}) = 8{,}473 \text{ ft}^3/\text{d at } 37°\text{C}.$$

12.10 Determine equivalent population:

$$\frac{13{,}460 \text{ lb/d}}{(0.5)(0.2 \text{ lb/capita}\cdot\text{d})} = 134{,}600 \text{ people}$$

Note: 0.5 = 50% solids reduction.

Required area at 2.0 ft^2 drying bed/capita:

$$(2.0)(134{,}600) = 269{,}200 \text{ ft}^2$$

Assume 270,000 ft^2 for a working estimate. Check capacity based on climate:

At 6% solids and 8 inch depth:

$$(270{,}000 \text{ ft}^2)(0.67 \text{ ft})(7.48 \text{ gal/ft}^2) = 1{,}353{,}132 \text{ gal}$$

Sludge volume accumulation:

$$(13{,}400 \text{ lb/d})\left(\frac{94\% \text{ water}}{6\% \text{ solids}}\right)\left(\frac{1}{8.34 \text{ lb/gal}}\right) = 25{,}284 \text{ gal/d of sludge}$$

Number of days of capacity:

$$\frac{1{,}353{,}132 \text{ gal}}{25{,}284 \text{ gal/d}} = 53.52 \text{ or } 54 \text{ d capacity}$$

Loss due to drainage:

At 6% solids: (0.94)(8 inches) – 7.52 inches of water

Note: 0.94 –94% water; 8 inches – placement depth given as an assumption.

Assuming the sludge drains to 18% solids, the depth of sludge plus water is

$$\frac{0.06}{0.18}(8 \text{ inches}) = 2.67 \text{ inches}$$

Of the 2.67 inches, (1.00 – 0.18)(2.67) = 2.19 inches is water. Therefore, anticipated loss due to drainage is 7.52 – 2.19 = 5.33 inches of water.

For the sludge to achieve 50% solids, the depth of sludge plus water is

$$\frac{0.06}{0.50}(8) = 0.96 \text{ inches}$$

Of the 0.96 inches, 50% is water, so the required evaporation is

$$2.19 - 0.48 = 1.71 \text{ inches}$$

The actual moisture balance is estimated as follows:
Calculate rainfall absorbed: (0.57)(40 inches/y) = 23 inches/y
The estimated evaporation rate: (0.75)(60 inches/y) = 45 inches/y
The net evaporation: 45 – 23 = 22 inches/y

Note: 0.57 = assumed absorption of precipitation; 0.75 = assumed evaporation potential.

Check the bed cycle rate:

$$\text{Beds can be cycled at:} \quad \frac{22 \text{ inches}}{1.71 \text{ inches}} = 13 \text{ times / y} \quad \text{or}$$

$$\frac{365 \text{ d / y}}{13 \text{ times / y}} = 28 \text{ d}$$

for a drying cycle. Since the bed capacity is 54 d, there is a safety factor of

$$\frac{54 \text{ d}}{28 \text{ d}} = 1.9$$

All of this assumes the annual average precipitation and evaporation occurs over the whole year. This is not likely. More refined analysis requires monthly moisture balance and allowance for storage. Covering the sand beds in winter helps tremendously.

Multiple Choice Problems

12.11 **b)** Narrow streams have less surface area than wide shallow ones and oxygen transfer from the atmosphere is limited by the reduced surface area in narrow streams. The depth of the stream reduces the diffusion of oxygen to the lower reaches.

12.12 **c)** Combined sewers carry both storm water and domestic sewage.

12.13 **c)** The recommended loading rate for primary settling tanks is 190 m³/d·m of weir length. The equivalent weir length for 0.100 m³/s of wastewater would be

$$\frac{\left(0.100 \text{ m}^3/\text{s}\right)\left(86\,400 \text{ s/d}\right)}{190 \text{ m}^3/\text{d} \cdot \text{m}} = 45.47 \text{ or } 45 \text{ m}$$

12.14 **d)** Settling design is controlled by overflow rate.

12.15 **a)** F/M ratio is defined as

$$\frac{F}{M} = \frac{QS_o}{VX} = \frac{\text{mg BOD}_5/\text{d}}{\text{mg MLVSS}} = \text{d}^{-1}$$

The influent BOD to the plant is considered to be a "constant," that is, it cannot be changed by the operator. The mass of microorganisms (MLVSS) can be regulated by the operator by adjusting the amount of sludge wasting. Greater sludge wasting will lower "M" in the F/M ratio.

12.16 **c)** A high rate trickling filter will have a hydraulic loading rate of 0.16 to 0.64 gal/ft²·min.

12.17 **b)** Nitrogen must be supplied in the ratio of 1:32 (N:BOD₅). For 800 mg/L BOD equivalent, the amount of nitrogen would be

$$N = \frac{800 \text{ mg / L}}{32} = 25 \text{ mg / L}$$

12.18 **e)** A "pinpoint" floc is indicative of a bulking sludge. This results from under aeration, low pH and low nitrogen and phosphorus.

12.19 **a)** Phosphorus is removed by chemical precipitation. Common precipitants are ferric chloride, alum and lime.

12.20 **e)** The diaphragm filter press yields the highest cake solids.

Highway Design

by Thomas L. Maleck

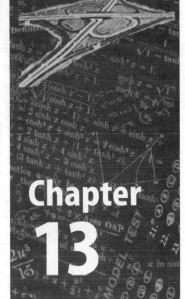

Chapter 13

Solutions to Practice Problems

13.1   **b)** 1700 pcphpl from Figure 13.1.

13.2   **d)** For a design speed of 60 mph, Detection and Recognition is expected to require from 2.0 to 3.0 seconds; Decision and Response Initiation is expected to require from 4.7 to 7.0 seconds; and the Maneuver is assumed to require 4.5 seconds:

> Minimum Decision Sight Distance = 2.0 + 4.7 + 4.5 = 11.2 sec
> Maximum Decision Sight Distance = 3.0 + 7.0 + 4.5 = 14.5 sec

Values from Table 13.4 have been used.

13.3   **c)** The others are wrong because of the following:

a) 6 lanes and a 10 foot shoulder are desirable for the high volumes. The median should be at least 40 feet wide.

b) 6 lanes and a 10 foot shoulder are desirable for the high volumes.

d) The cross slope of the pavement should be at least 1.5%.

13.4   Analyze each interval.

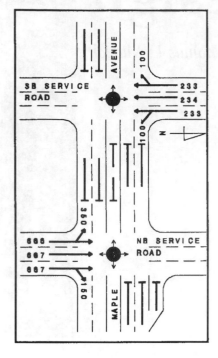

Figure 13.34

Interval 1:

Interval 1 is illustrated in Figure 13.34. During this interval of time the north and southbound movements have the green light and all others are stopped. Traffic assumed to be moving in this interval are assigned to the appropriate lanes. The movement which will require the most time is the northbound movement. Using Eq. 13.1.1 the time is

$$t = \frac{2.1M_c}{N_c} + 3.7$$

$$= \frac{2.1(667)}{45} + 3.7 = 34.8 \text{ sec}$$

where M_c = 667 vehicles/hr and N_c = 3600/80 = 45 cycles/hr.

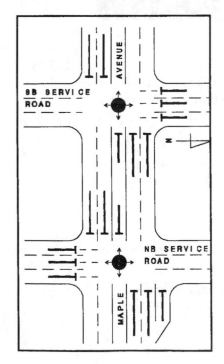

Figure 13.35

Interval 2:

Interval 2 is illustrated in Figure 13.35. This is the amber or clearance time. The amber time is a function of the approach speed which is assumed to be 30 mph. From Eq. 13.1.2 the required time for this interval is

$$Y = t_r + \frac{V}{2a + 64.4g} + \frac{W + L}{V}$$

$$= 1 + \frac{44}{2 \times 14} + \frac{72 + 20}{44} = 4.7 \text{ sec}$$

where we have assumed t_r = 1 sec, a = deceleration = 14 ft/sec^2, V = 44 ft/sec (30 mph speed limit), and g = 0. With six lanes at 12 ft, W = intersection width = 72 ft and L = length of vehicle = 20 ft.

Interval 3:

Interval 3 is illustrated in Figure 13.36. This interval is for a lead green to clear the left-turn vehicles (from the service roads) and prevent their interfering with the east and westbound through movements on Maple Avenue. The time required for this interval is equal to the time needed to start the vehicles waiting on Maple Avenue between the two service drives less the travel time for the through traffic arriving on Maple Avenue.

$$\frac{175 \text{ vph}}{45 \text{ cycles/hour}} = 3.9 \text{ Vehicles per Cycle.}$$

Since this is less than 5 vehicles, Eq. 13.1.1 can not be used. Hence, assume 4 vehicles:

Clearance Time = 3.8 + 3.1 + 2.7 + 2.4 = 12.0 sec.

Travel Time = 260 ft/66 ft/sec = 3.9 sec

Time Required = 12.0 − 3.9 = 8.1 sec

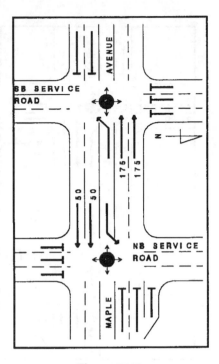

Figure 13.36

Interval 4:

Interval 4 is illustrated in Figure 13.37. During this interval the east and westbound movements occur. The westbound movement is critical because it requires the most time. Using Eq. 13.1.1 the required time for this movement is

$$t = \frac{2.1 \times 400}{45} + 3.7 = 22.4 \text{ sec}$$

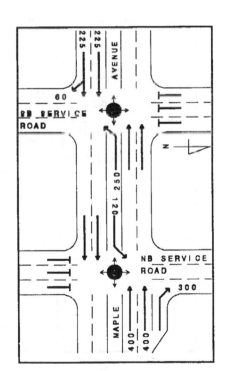

Figure 13.37

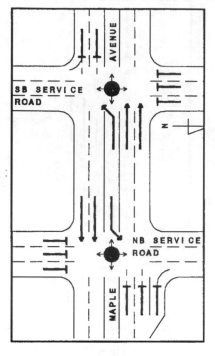

Figure 13.38

Interval 5:

Interval 5 is illustrated in Figure 13.38. This interval is the amber time for the east and westbound movements. From Eq. 13.1.2 this time is

$$Y = t_r + \frac{V}{2a + 64.4g} + \frac{W + L}{V}$$

$$= 1 + \frac{66}{2 \times 14} + \frac{36 + 20}{66} = 4.2 \text{ sec}$$

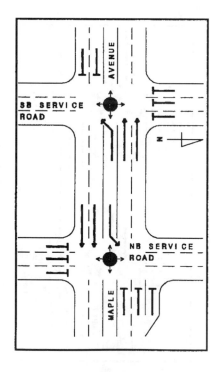

Figure 13.39

Interval 6:

Interval 6 is illustrated in Figure 13.39. This interval is to allow the last vehicle to clear the amber in interval 5 to also clear the amber in interval 7. The time period is a function of the speed limit and the distance to be traveled less the amber time in interval 7:

Clearance Time = 260 ft/ 66 ft/sec = 3.9 sec
Amber Time = 4.2 sec
Time required = 3.9 − 4.2 = −0.3 sec

A negative value means that there will be a time period of 0.3 seconds when both Ambers will be on.

Interval 7:

This interval is illustrated in Figure 13.40. It has the same length of time as in interval 5:

$$Y = 4.2 \text{ sec}$$

The capacity of this geometric configuration is ratio of the time required to accommodate the demand with respect to the total time available. The time required is the summation of the critical times for each interval:

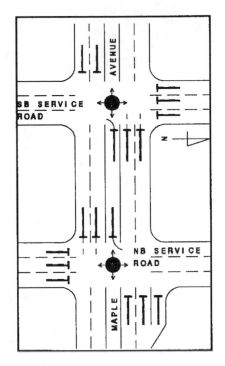

Interval 1:	34.8
Interval 2:	4.7
Interval 3:	8.1
Interval 4:	22.4
Interval 5:	4.2
Interval 6:	−0.3
Interval 7:	4.2
Total	78.1 sec

Time Available = 80 sec

Capacity = 78.1/80 = 0.98 or 98%

The volume-to-capacity ratio is between 0.9 and 1.0. Therefore the Level-of-Service is "E".

Figure 13.40

13.5   Use Eq. 13.2.2:   Total number of accidents = Injury + PDO = 293 ADT = 14600

Number of fatal accidents = 6   Number of injury accidents = 89

$$R \text{ (all accidents)} = \frac{293 \times 1,000,000}{14600 \times 4 \times 365} = 13.7 \text{ MEV}$$

$$R \text{ (fatal accidents)} = \frac{6 \times 1,000,000}{14600 \times 4 \times 365} = 0.28 \text{ MEV}$$

$$R \text{ (injury accidents)} = \frac{89 \times 1,000,000}{14600 \times 4 \times 365} = 4.18 \text{ MEV}$$

$$\text{Severity Ratio} = \frac{89}{293} = 0.30$$

13.6   A) Determine R and T as a function of V using Eq. 13.4.1:

$$R = \frac{V^2}{(e+f)15} = \frac{V^2}{(0.06+0.06)15} = \frac{V^2}{1.8}$$

Tangent leg of Curve #1:

$$T_1 = R \tan \frac{\Delta_1}{2} = \frac{V^2}{1.8}(0.2679) = 0.14886V^2$$

Tangent leg of Curve #2:

$$T_2 = \frac{V^2}{1.8}(0.57705) = 0.320V^2$$

Tangent distance t for a 3 sec transition between curves is

$$t = (3)(1.47)V = 4.41V$$

Total available distance is 1000 feet. Therefore,

$$1000 = t + T_2 + T_1$$

$$1000 = 4.41V + 0.320V^2 + 0.149V^2$$

$$V^2 + 9.39V - 2129 = 0$$

$$\therefore V = \frac{-9.39 \pm \sqrt{9.39^2 - 4(-2129)}}{2} = 41.7 \text{ mph}$$

B) Curve #1

$$R = 966.001$$

$$D = 5729.6/966 = 5.93°$$

$$T_1 = R\tan\left(\tfrac{\Delta_1}{2}\right) = 258.89$$

$$E_1 = T_1\tan\left(\tfrac{\Delta_1}{4}\right) = 34.08$$

$$M_1 = E_1\cos\left(\tfrac{\Delta_1}{2}\right) = 32.92$$

$$L_1 = 100(\Delta_1/D) = 505.90$$

Curve #2

$$T_2 = R\tan\left(\tfrac{\Delta_2}{2}\right) = 557.72$$

$$E_2 = T_2\tan\left(\tfrac{\Delta_2}{4}\right) = 149.44$$

$$M_2 = E_2\cos\left(\tfrac{\Delta_2}{2}\right) = 129.42$$

$$L_2 = 100(\Delta_2/D) = 1011.80$$

C) Stationing

PI_1	=	*132 + 50.00*
$-T_1$	=	*2 + 58.84*
PC_1	=	*129 + 91.16*
$+L_1$	=	*5 + 05.90*
PT_1	=	*134 + 97.06*
$+t$	=	*1 + 83.44*
PC_2	=	*136 + 80.50*
$+L_2$	=	*10 + 11.80*
PT_2	=	*146 + 92.30*

13.7   The given data and assumptions are

$h_1 = 3.5$ ft
$h_2 = 4.25$ ft
$S = 1200$ ft
$A = 3 - 0 = 3$

For $S < L$ use Eq. 13.5.3:

$$L = \frac{3 \times 1200^2}{100\left(\sqrt{7.0} + \sqrt{8.50}\right)^2} = 1397 \text{ ft}$$

Since $S < L$, this is the correct answer.

Check by solving the other possibility $S > L$ using Eq. 13.5.4:

$$L = 2 \times 1200 - \frac{200 \times \left(\sqrt{3.5} + \sqrt{4.25}\right)^2}{3} = 1369 \text{ ft}$$

This gives $S < L$, which is not acceptable.

13.8   The sight distance requirement for a right turn maneuver is only one to three feet less than required with the sight distance from the right for a left turn maneuver onto a 2-lane 2-way highway. Therefore the procedure and answer is essentially the same as in Example 13.17. $\therefore d = 654$ ft.

13.9   A) From Table 13.7, the largest vehicle is a WB-40 which only needs a 150-ft radius. The next largest truck which is a WB-50, needs at least a 200-ft radius.

B) From Table 13.7, the largest truck which is a WB-62, only needs a 120-ft radius. Therefore all of the trucks listed can make this turn.

C) No, the radius will not have to be increased. From Table 13.7, the minimum radius actually decreases from 120 to 115 ft. Even when the angle is increased to 180°, the radius still drops to 55 ft.

D) No, as can be seen in Table 13.7 for a 180° turn. The SU truck needs a radius of 30 ft while the WB-50 needs a radius of only 25 ft.

13.10   The first step is to draw a space-time diagram that matches the distances and times required for this problem, as in Fig. 13.41.

A) Fig. 13.41 shows the space-time diagram with the given information.

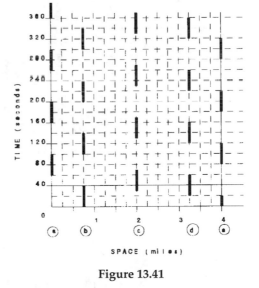

Figure 13.41

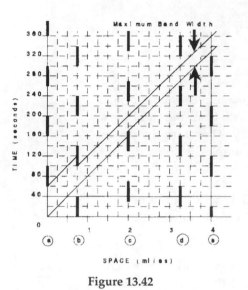

Figure 13.42

B)   By drawing parallel lines representing 45 mph, the band width can be determined. Fig. 13.42 shows the maximum band width for the existing situation as 30 sec.

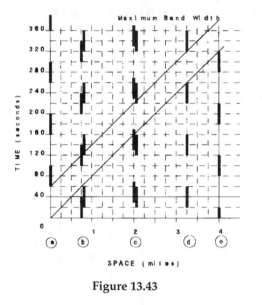

Figure 13.43

C)   By changing the offsets at signal "b" and signal "c" (to 60 and 20 sec, respectively), the band width can be increased to 60 sec. This is shown in Fig. 13.43. This is the maximum possible band width because signal "a" has an effective green equal to 60 sec. Since the band width can never increase, 60 sec is the widest band width.

13.11   Step 1: The initial time of concentration is assumed to be 15 minutes, and $i = 3.8$ in/hr. Therefore the estimated runoff expected to reach Catch Basin A is

$$Q_1 = ciA = 0.35 \times 3.8 \times 1.75 = 2.33 \text{ cfs}$$

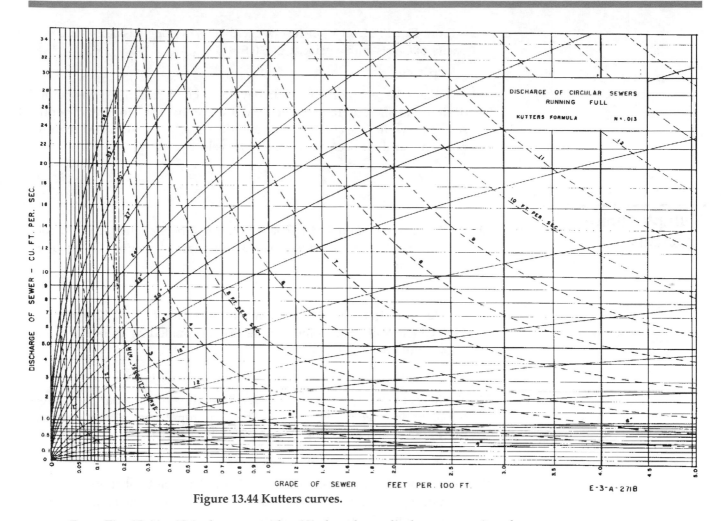

Figure 13.44 Kutters curves.

From Fig. 13.44 a 12-inch sewer with a 1% slope has a discharge capacity of 3.6 cfs with a velocity of 4.5 fps. The time for the flow to reach Catch Basin B is

$$275 \text{ ft} / 4.5 \,(\text{ft}/\text{sec}) = 61 \sec = 1 \min$$

Step 2: The time of concentration is now 15 + 1 = 16 min, and i = 3.7 in/hr. Therefore the estimated runoff expected to reach Catch Basin B is

$$Q_2 = ciA = 0.30 \times 3.7 \times 1.90 = 2.11 \text{ cfs}$$

resulting in

$$Q_1 + Q_2 = 2.33 + 2.11 = 4.44 \text{ cfs}$$

From Fig. 13.44 a 15-inch sewer with a 0.8% slope has a discharge capacity of 5.8 cfs and a velocity of 4.7 fps. The time required for the flow to reach Catch Basin C is

$$275 \text{ ft}/(4.7 \text{ ft}/\text{sec}) = 59 \text{ sec} = 1 \min$$

Step 3: The time of concentration is now 17 minutes with i = 3.6 in/hr. Therefore, the estimated runoff expected to reach Catch Basin C is

$$Q_3 = ciA = 0.45 \times 3.6 \times 1.50 = 2.43 \text{ cfs}$$

and

$$Q_1 + Q_2 + Q_3 = 4.44 + 2.43 = 6.87 \text{ cfs}$$

From Fig. 13.44, an 18-inch sewer with a 0.7% slope has a discharge capacity of 8.8 cfs with a velocity of 5.0 fps.

Step 4: etc.

It may take several iterations to develop a good design. To this point the design has an acceptable velocity and capacity. The slope and depth of sewer are within the critical values and the velocity has increased slightly; at no point has there been a decrease in the velocity.

Multiple Choice Problems

13.12   **d)**   Refer to equation 13.4.1. The maximum superelevation, e, for a cold climate should not exceed 0.08 and with no lateral friction, f is equal to 0.00.

13.13   **c)**   Refer to equation 13.4.1. The maximum value of f is 0.12 (page 154, 1990 Green Book). The minimum radius is provided by using the maximum values of e and f.

13.14   **c)**   Refer to Equations 13.4.4 and 13.4.10. With perpendicular legs the external angle, Δ, equals 90 degrees. The minimum length of curve also requires the minimum radius. Therefore $L = 1884.96$ ft.

13.15   **d)**   From page 178 of the 1990 Green Book. The length of runoff is 180 ft. for a design speed of 60 mph, $e = 0.08$ and 10 ft. lanes. The tangent section is to contain 60 to 80% of the runoff.

13.16   **e)**   All of the above are correct. However, d) is the recommended practical control for length of spiral. (Page 174 of the 1990 Green Book).

13.17   **c)**   A 10 year storm is adequate for a county highway. There is 2.91 acres of pavement and a runoff coefficient of 0.85 was assumed. There is 0.73 acres of gravel surface with an assumed runoff coefficient of 0.50. There is 4.36 acres of turf right-of-way with an assumed coefficient of runoff of 0.40. For a 10 year storm and an initial time of concentration of 15 minutes the runoff from Figure 13.31 is estimated as 3.8 inches per hour. It was assumed that little if any significant runoff would come from the forest and the only runoff would be generated from within the limits of the right-of-way.

13.18   **a)**   The most efficient ditch would have the lowest value of Manning's coefficient and have the smallest cross-sectional area. For this case the design flow is 8.7 cfs. (assuming a ditch on each side). The slope is 132/5280 ft/ft. By using Equation 13.9.2, the resulting cross-section area is 1.445 sq ft. The wetted perimeter is 3.8 ft. The resulting flow is 8.9 cfs with a velocity of 6.2 fps.

13.19　**e)**　The grass lining has the highest Manning's coefficient and the 4-ft flat bottom ditch produces the lowest Hydraulic radius (R). These two factors increase the cross-section area and decrease the rate of flow for the same runoff.

13.20　**a)**　From Kutters Curves the required grade of the sewer for answer a) is 2.5% which is appropriate for this location.

Soil Mechanics

by Thomas F. Wolff

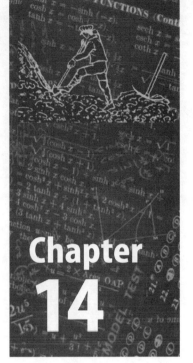

Chapter

14

Solutions to Practice Problems

14.1 **b)** The weight of moist soil is $48.30 - 7.41 = 40.89$ g

The weight of dry soil is $41.22 - 7.41 = 33.81$ g

The weight of water is $40.89 - 33.81 = 7.08$ g

The water content is $7.08 / 33.81 = 0.209$ or 20.9 percent.

14.2 **a)** As the quantity of soil is not or known, and only relationships are of interest, assume the total volume V is 1.00 ft³. The total weight W is then 120 lb. As the soil is saturated, the volume of air $V_a = 0$. After sketching a phase diagram, two simultaneous equations can be written, one on the volume side and one on the weight side:

$$V_s + V_w = 1.00$$

$$W_s + W_w = 120.0$$

W_s can be written as $G_s \gamma_w V_s = (2.70)(62.4) V_s$ and W_w can be written as $\gamma_w V_w = 62.4 \, V_w$

Hence, there are two simultaneous equations

$$V_s + V_w = 1.00$$

Multiple Choice Problems

$$168.4\ V_s + 62.4V_w = 120.0$$

The solution is

$$V_s = 0.543,\ \ V_w = 0.456$$

Then

$$W_s = (2.70)(62.4)(0.543) = 91.5\ \text{lb}$$

The required dry unit weight is

$$\gamma_d = W_s/V = 91.5/1.00 = 91.5\ \text{lb/ft}^3$$

14.3 **c)** Use the relationship $Se = wG_s$. Then

$$e = wG_s/S$$

$$e = (0.20)(2.75)/1 = 0.55$$

14.4 **a)** Since 10 percent of the sample passes the No. 200 sieve (the *fines*), 90 percent is larger than the No. 200 sieve, and the soil is coarse grained. The first letter of the classification must be *S* for sand or *G* for gravel.

70 percent of the total sample passed the No. 4 sieve and 30 percent was larger than the No. 4 sieve. The sample is 30 percent gravel and 10 percent fines, leaving 60 percent sand. The *coarse fraction* is the 90 percent larger than the No. 200. The coarse fraction is 30/90 = 33% gravel and 60/90 = 67% sand. Since the coarse fraction contains more sand than gravel, the first letter of the classification is *S*.

As the percent passing the No. 200 sieve is between 5 and 12, the sample requires a dual classification, with the first symbol being *SP* or *SW*, and the second being *SC* or *SM*.

The coefficient of uniformity must be greater than 6 for the soil to be a well-graded *SW*; as it is not, the first symbol is *SP*.

As the soil is stated to be non-plastic, the fines must be silt rather than clay; the second symbol is *SM*.

The soil is classified as *SP-SM*.

14.5 **b)** If $LL = 45$ and $PL = 30$, then $PI = 45 - 30 = 15$. Referring to the coordinates (45,15) on the plasticity chart, the soil is either *ML* or *OL*. If it is not stated that the soil is significantly organic, it would be classified as *ML*.

14.6 **b)** Using equation 14.9.3, the relative density is:

$$[(1/105) - (1/121)]/[(1/105) - (1/125)] = 0.827\ \text{or}\ 82.7\%$$

Calculations involving ratios of differences are numerically sensitive. To obtain three digit accuracy in the solution, four or more digits must be carried through the calculations.

14.7   **d)**   The total vertical stress is

$$\sigma = (2.0)(20.0) = 40.0 \text{ kN/m}^2$$

The pore water pressure is

$$u = (2.0)(9.81) \qquad = 19.6 \text{ kN/m}^2$$

The effective vertical stress is

$$\sigma' = 40.0 - 19.6 = 20.4 \text{ kN/m}^2$$

14.8   **b)**   As previously shown, the total vertical stress is 40.00 kN/m².

14.9   **a)**   The horizontal effective stress is K times the vertical effective stress:

$$\sigma'_h = (0.5)(20.4) = 10.2 \text{ kN/m}^2$$

14.10   **b)**   The total head at B is 100 cm. The total head at C is 80 cm. The head loss from B to C is

100 - 80 = 20 cm. The 20 cm head loss occurs over a length of 80 cm.

For the uniform flow conditions shown, the hydraulic gradient between B and C is the head loss divided by the length of seepage path. Thus

$$i = 20/80 = 0.25$$

14.11   **a)**   At point B, the total head is 100 cm. The elevation head is 10 cm. Thus, the pressure head is

90 cm or 0.90 m. The water pressure is the pressure head times the unit weight of water, or

$$u = (0.90 \text{ m})(9.81 \text{ kN/m}^3) = 8.83 \text{ kN/m}^2$$

14.12   **d)**   From the flow net, three of the six equipotential drops have been crossed where the seepage path reaches the tip of the sheetpile. Of the total head loss of 100 - 90 = 10 ft, one half (or 5 ft) has occurred by this point. The total head at the tip of the sheetpile is 95 ft.

14.13   **c)**   From problem 14.12, the total head is 95 ft. The elevation head is 79 ft. The pressure head is

95 - 79 = 16 ft. The pore water pressure is the pressure head time the unit weight of water

$$(16)(62.4) = 998 \text{ lb/ft}^2$$

14.14 **a)** An overconsolidated soil is one where the preconsolidation pressure is greater than the effective overburden pressure.

14.15 **e)** For saturated clay tested in undrained shear, the friction angle is zero and = c. For an unconfined compression test, $\sigma_3 = 0$. Since $c = (\sigma_1 + \sigma_3)/2$, the compressive stress required for failure is $\sigma_1 = (2)(400) = 800$ lb/ft^2. The area of the sample to which the compressive force is applied is $\pi(0.5)^2/4 = 0.1964$ ft^2. The force required to cause shear failure is

$$F = \sigma A = (800)(0.1964) = 157 \text{ lb}$$

Essay Problems

14.16   The mass of water M_w is

$$46.95 - 35.06 = 11.89 \text{ g or } 0.01189 \text{ kg}$$

The water content is

$$M_w/M_s = 11.89/35.06 = 0.339 \text{ or } 33.9 \text{ percent}$$

The volume of solids is

$$V_s = M_s/G_{sw} = 0.03506 \text{ kg}/(2.70)(1000 \text{ kg/m}^3) = 13.0 \times 10^{-6} \text{ m}^3$$

The volume of water is

$$V_w = M_w/\gamma_w = 0.01189/1000 = 11.9 \times 10^{-6} \text{ m}^3$$

As the soil is saturated, $V_A = 0$.

The void ratio is

$$e = V_v/V_s = 11.9 \times 10^{-6}/13.0 \times 10^{-6} = 0.915$$

The porosity is

$$n = V_v/V = 11.9 \times 10^{-6}/24.9 \times 10^{-6} = 0.478$$

The total unit weight is

$$\gamma = W_t/V = Mg/V = (0.04695 \text{ kg})(9.81 \text{ m/s}^2)/24.9 \times 10^{-6} \text{ m}^3$$

$$= 18\ 500 \text{ kg/s}^2 \cdot \text{m}^2$$

$$= 18.5 \text{ kN/m}^3$$

The dry unit weight is

$$\gamma_d = W_s/V = M_s g/V = (0.03506 \text{ kg})(9.81 \text{ m/s}^2)/24.9 \times 10^{-6} \text{ m}^3$$

$$= 13\,800 \text{ kg/s}^2 \cdot \text{m}^2$$

$$= 13.8 \text{ kN/m}^3$$

14.17   As the quantity of soil is not fixed, assume $W_s = 100$ lb. Then

$$W_w = 20 \text{ lb},$$

V_s	$= 100 /(2.68)(62.4)$	$= 0.598 \text{ ft}^3$
V_w	$= 20 /62.4$	$= 0.321 \text{ ft}^3$
V_v	$= 0.321 /0.80$	$= 0.401 \text{ ft}^3$
V	$= 0.598 + 0.401 + 0$	$= 0.999 \text{ ft}^3$

The required quantities are

$$\gamma = (100 + 20)/(0.598 + 0.401) = 120.1 \text{ lb/ft}^3, \text{ or}$$

$$120.1 \,(9.81/62.4) = 18.9 \text{ kN/m}^3$$

$$\gamma_d = 100/(0.598 + 0.321) = 108.8 \text{ lb/ft}^3, \text{ or}$$

$$100.1 \,(9.81/62.4) = 15.7 \text{ kN/m}^3$$

14.18   Plot the dry unit weight vs water content:

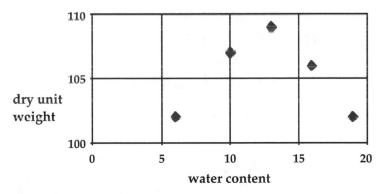

The maximum dry unit weight is approximately 109.0 lb/ft³. The optimum water content, the *x* value, corresponding to the peak of the curve, is 13 percent. The soil is to be compacted to at least 95 percent of maximum dry unit weight, or 103.6 lb/ft³. Extending a line at 103.6, it is seen that the minimum and maximum water contents for which the minimum specified density can be achieved are about 7 percent and 18 percent.

14. 19 From 5 lbs of moist soil with a water content of 20 percent, one can write

$$\left.\begin{array}{l} W_s + W_w = 5.0 \\ W_s + 0.20W_s = 5.0 \end{array}\right\} \quad \therefore W_s = 4.17 \text{ lb}, \quad W_w = 0.83 \text{ lb}$$

To raise the water content from 20 percent to 30 percent, an additional 10 percent by weight of water must be added, or $(0.10)(4.17) = 0.42$ lb.

14.20 The required minimum weight of soil solids in the fill is $(20)(10\ 000) = 2.0 \times 10^5$ kN. As the soil in the borrow area has a dry unit weight of only 19.0 kN/m³, the required volume of excavation is

$$2.0 \times 10^5 \text{ kN}/19.0 \text{ kN/m}^3 = 10530 \text{ m}^3$$

The required weight of water in the fill is $(0.10)(2.0 \times 10^5) = 20\ 000$ kN. The weight of water that will come with the borrow soil when hauled is $(0.08)(2.0 \times 10^5) = 16\ 000$ kN. The required water to be added is 20 000 - 16 000 = 4000 kN, or $(4000 \text{ kN})/(9.81 \text{ kN/m}^3) = 408$ m³ or 4.08×10^5 liters.

14.21 At the ground surface, all quantities are zero.

At 4 ft:

$$\sigma_v \quad = (4)(125) = 500 \text{ lb/ft}^2$$
$$u \quad = 0 \text{ (above water table)}$$
$$\sigma'_v \quad = 500 - 0 = 500 \text{ lb/ft}^2$$
$$\sigma'_h \quad = (500)(0.40) = 200 \text{ lb/ft}^2$$
$$\sigma_h \quad = 200 + 0 = 200 \text{ lb/ft}^2$$

At 10 ft:

$$\sigma_v \quad = 500 + (6)(130) = 1280 \text{ lb/ft}^2$$
$$u \quad = (6)(62.4) = 374 \text{ lb/ft}^2$$
$$\sigma'_v \quad = 1280 - 374 = 906 \text{ lb/ft}^2$$
$$\sigma'_h \quad = (0.40)(906) = 362 \text{ lb/ft}^2$$
$$\sigma_h \quad = 362 + 374 = 736 \text{ lb/ft}^2$$

14.22 The *e*-log *p* curve is plotted below.

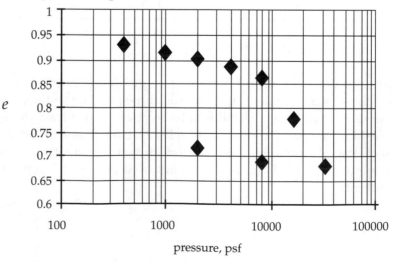

The preconsolidation pressure is approximately 9000 psf.

The compression index is slope of the virgin curve (to the right of 9000 psf). Extending the curve through the points (10 000, 0.85) and (60 000, 0.6) and changing the sign to positive the compression index is

$$C_c \quad = -(0.85\text{-}0.60)\,/(\log 10\,000 - \log 60\,000) = 0.321$$

The recompression index is obtained by extending a line through the unloading curve through the points (0.073, 1000) and (0.069, 10 000). The recompression (or swell) index is:

$$C_s \quad = -(0.073 - 0.069)\,/(\log 1000 - \log 10\,000) = 0.04$$

14.23   As the clay later is being unloaded, the recompression index (or swell index) is used.

The change in void ratio due to reducing the effective vertical stress from 1800 to 1300 psf is

$$\Delta e \quad - C_r \log \left[(p + \Delta p)/p\right]$$

$$= 0.040 \log (1800/1300) = 0.00565$$

As the clay is swelling due to unloading, this is an increase in void ratio.

The average vertical strain is estimated using the midpoint void ratio change as

$$\varepsilon_v \quad = \Delta e \,/(1 + e_0)$$

$$= 0.00565/1.900 = 0.00298$$

The heave is the average strain times the thickness of the layer:

$$\Delta H \quad = (0.00298)(5.0)$$

$$= 0.0149 \text{ ft}$$

$$= 0.18 \text{ inches}$$

14.24   Draw two Mohr's circles for the given data and read off the strength as $c = 0.40$ tsf, $\phi = 19° \pm$

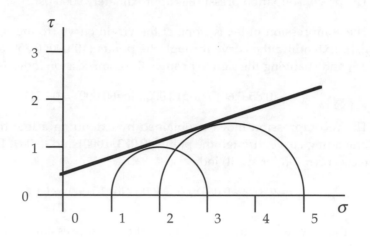

14.25   With 100 lb shear force applied, the shear stress on the sample is

$$\tau = 100/3^2 = 11.11 \text{ psi}$$

Use the Mohr-Coulomb failure equation:

$$\tau = c + \sigma \tan \phi$$

$$11.11 = 0 + \sigma \tan 30. \quad \therefore \quad \sigma = 19.25 \text{ psi}$$

The maximum vertical force that can be used is (19.25 psi) (3^2 in^2) = 173 lb

---

Multiple Choice Problems (PE Format)

14.26 **b)** From Equation 14.5.5,

$$\gamma_{sat} = \frac{(G_s + Se)\gamma_w}{1+e} = \frac{[2.75+1.0(0.5)]62.4}{1.5} = 135.2 \text{ lb/ft}^3$$

14.27 **c)** The excavation is symmetrical about the centerline, which is a flow line equivalent to an impervious boundary. Each of the two sides can be considered a separate flow net with N_f/N_e = 2/6. From Equation 14.11.6:

$$Q = kH \frac{N_f}{N_e} \times 2 = (0.1 \text{ ft/min})(12 \text{ ft})\left(\frac{2}{6}\right)(2 \text{ sides}) = 0.8 \text{ ft}^3/\text{min/ft}$$

This is the flow per lineal foot of excavation. For a 100 ft long excavation,
$$Q_{total} = (0.8 \text{ ft}^3/\text{min/ft})(100 \text{ ft}) = 80 \text{ ft}^3/\text{min}$$

14.28 **d)** Point A is 8 ft below the surface of a soil with a total (saturated) unit weight of 135.2 lb/ft^3. From Equation 14.10.1,
$$\sigma_v = \gamma_{sat} H = (135.2 \text{ lb/ft}^3)(8.0 \text{ ft}) = 1081.6 \text{ lb/ft}^3$$

14.29 **a)** As water is flowing, the pore pressure must be determined from a seepage analysis. As 12 ft of total head is lost over six equipotential drops, each equipotential drop (square) represents 2 ft of total head loss. Point A is one drop above the seepage exit, hence the total head at Point A is:

$$TH = 388 + (1/6)(12) = 390 \text{ ft}$$

or

$$TH = 400 - (5/6)(12) = 390 \text{ ft}$$

The elevation head at A is 380. The pressure head is the total head minus the elevation head, or

$$PH = 390 - 380 = 10 \text{ ft}$$

The pore pressure at A is then

$$u = (PH)\, \gamma_w = (10 \text{ ft})(62.4 \text{ lb/ft}^3) = 624 \text{ lb/ft}^2$$

14.30 **e)** From Equation 14.10.4

$$\sigma'_v = \sigma_v - u = 1081.6 - 624.0 = 457.6 \text{ lb/ft}^2$$

Foundations and Retaining Structures

by Thomas F. Wolff

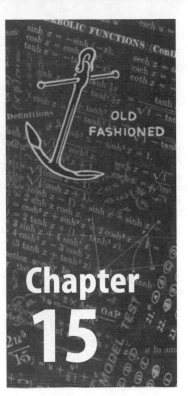

Chapter 15

Solutions to Practice Problems

15.1   **b)** The 4 blows for the first six inch increment are discarded and the blows for the second and third increment are added.

$$N = 6 + 8 = 14$$

Multiple Choice Problems

15.2   **c)** At a depth of 10 feet, the total vertical stress is

$$\sigma_v = 10(120) = 1200 \text{ psf}$$

As the water table is at a depth of 15 ft, the pore pressure at a depth of 10 ft is zero, and the effective stress σ_v' is also equal to 1200 psf. To calculate the correction using Equation 15.1.1, the effective overburden stress must be converted to tons per square foot (1200 psf = 0.6 tsf). Then

$$C = \sqrt{\frac{1}{\sigma_v'}} = \sqrt{\frac{1}{0.6}} = 1.29$$

15.3 **a)** From Equation 14.15.1, the cohesion is one-half the undrained strength; hence $c = 1000$ psf. From Equation 15.4.7, the net ultimate bearing capacity is

$$q_{ult} = 1000(5.14) = 5140 \text{ psf}$$

15.4 **c)** As sand is cohesionless, the first term of the bearing capacity equation will be zero. From Table 15.1, $N_q = 23.2$ and $N_\gamma = 30.2$. From Equation 15.4.9, the shape factors are

$$F_{qs} = 1 + (3/3)\tan 30° = 1.577$$

$$F_{\gamma s} = 1 - 0.4(3/3) = 0.6$$

From Equation 15.4.8, the ultimate bearing capacity is

$$q_{ult} = 0 + 125(0)(1.577) + 0.5(3)(125)(30.2)(0.6) = 3400 \text{ psf}$$

Multiplying the ultimate bearing capacity by the area of the footing, the ultimate force to cause failure in bearing is

$$Q_{ult} = 3400 \ (3^2) = 30,600 \text{ lb}$$

15.5 **a)** Vertical stress increase due to a point load is calculated using Equation 15.5.1 or 15.5.2. For a point 5 ft directly below the load, $R = z = 5$ ft. Then

$$q_v = 3 \ (1000) \ /[\ 2\pi \ (5^2)\ ] = 19 \text{ psf}$$

15.6 **b)** The applied pressure at the base of the footing is

$$q_o = p \ /BL = 10,000 \ /4(4) = 625 \text{ psf}$$

The average pressure on a plane 5 ft deep is calculated using Equation 15.5.7

$$q_v = 625(4)(4) \ /[(4 + 5)(4 + 5)] = 123 \text{ psf}$$

15.7 **b)** The settlement is calculated using Equation 15.6.2. First the influence factor α is calculated from Equation 15.6.3; from Example 15.12, $\alpha = 1.122$ for the center of square foundations. Then, from Equation 15.6.2,

$$S_e = [4(625) \ /200,000] \ (1 - 0.4^2) \ (1.122) = 0.0118 \text{ ft} = 0.14 \text{ in}$$

15.8 **d)** Refer to equations 15.8.1. As the layer is singly drained, $H = 20$. Fifty percent of the settlement will have occurred when consolidation is 50 percent complete. From Table 15.2, this occurs when $T = 0.197$. Then

$$t_{50} = T_{50} H^2 \ /c_v = 0.197(20^2) \ /0.4 = 197 \text{ days}$$

15.9 **c)** The undrained strength or cohesion c is one half the unconfined compressive strength, or 1000 psf. The cross sectional area of the pile tip is

$$A_{tip} = (14/12)^2 = 1.36 \text{ ft}^2$$

From Equation 15.7.4, the tip capacity is

$$Q_{tip} = 9 A_{tip} c = 9(1.36)(1000) = 12{,}240 \text{ lb} = 12.2 \text{ kips}$$

15.10 **a)** The cohesion is one-half the unconfined compressive strength, or 1000 psf. For short-term loading of clays, undrained conditions apply and $\phi = 0$; hence $K = 1$. From Example 15.19, the depth to zero horizontal pressure is

$$z = 2c / \gamma \ K^{0.5} = 2(1000) \ / (100)(1) = \mathbf{20} \text{ ft}$$

15.11 **b)** Using equation 15.8.3, for $\phi = 30°$, $K_a = 0.333$. From Example 15.16, the total active earth force is

$$P_a = \frac{1}{2} K_a \gamma H^2 = 0.5(0.333)(125)(10^2) = 2080 \text{ lb /ft}$$

15.12 As the foundation is clean sand, $c = 0$ and the first term of the bearing capacity equations will drop out. The bearing capacity factors are obtained from Table 15.1; for $\phi = 32°$, $N_q = 23.2$, $N_\gamma = 30.2$. The shape factors are obtained from Equations 15.4.9:

Essay Problems

$$Fq_s = 1 + (4 \ /4) \tan 32° = 1.62$$

$$F_{\gamma s} = 1 - 0.4 \ (4/4) = 0.6$$

As the water table is at the base of the footing, the total unit weight is used in the second term and the buoyant unit weight is used in the third term of Equation 15.4.8:

$$q_{ult} = 0 + (125)(3)(23.2)(1.62) + (0.5)(4)(125 - 62.4)(30.2)(0.6)$$
$$= 16{,}400 \text{ psf}$$

Using a factor of safety of 3.0, the allowable bearing pressure would be

$$16{,}400/3 = 5466 \text{ psf} \approx 5500 \text{ psf}$$

15.13 The bearing factors are the same as in the previous problem. For a continuous or strip footing, the shape factors are 1.0. From Equation 15.4.8, the ultimate bearing pressure is

$$q_{ult} = 0 + (125 - 62.4)(3)(23.2) + (0.5)(3)(125 - 62.4)(30.2)$$
$$= 7193 \text{ psf}$$

Using a factor of safety of 3.0, the allowable bearing pressure would be

$$7193 / 3 = 2397 \text{ psf} \quad \sim 2400 \text{ psf}$$

15.14   For foundation design in clay, the undrained ($\phi = 0$) conditions will govern. For $\phi = 0$, $N_c = 5.14$ and $N_q = 1$. For a continuous or strip footing, the shape factors are 1.0. From Equation 15.4.8, the ultimate bearing pressure is

$$q_{ult} = (1500)(5.14)(1.) + (125)(3)(1) = 8100 \text{ psf}$$

Taking the factor of safety as 3.0, the allowable bearing pressure is

$$q_a = 8100 / 3 = 2700 \text{ psf}$$

15.15   From Equation 15.4.8, the ultimate bearing capacity of a footing on the surface of sand is

$$q_{ult} = 0 + 0 + (.5)B\gamma N_\gamma (0.6)$$

For a second footing twice as wide, the B above would be replace by $2B$; thus the ultimate bearing pressure doubles when the footing width doubles. However, the second footing has four times the base area of the first footing. As it can carry twice as much force per unit area (pressure) over four times the area, the required force to fail the second footing is eight times the force required to fail the first footing.

15.16   Equation 15.6.2 predicts settlement of flexible footings in sand. From Section 15.6, the modulus can be estimated as $16N$ kips/ft^2, or

$$(16)(15) = 240 \text{ kips/ft}^2$$

Poisson's ratio μ can be taken as 0.4. From Example 15.12, $\alpha = 1.122$ for the center of a square footing. Setting S_e to 1 inch (0.0833 ft), and rearranging Equation 15.6.2, the bearing pressure for 0.0833 ft settlement is

$$q_o = (0.0833 \text{ ft})(240 \text{ kips/ft}^2) / [(4 \text{ ft})(1 - 0.4^2)(1.122)]$$
$$= 5.30 \text{ kips/ft}^2 = 5300 \text{ psf}$$

For settlement under the corner of a flexible footing, α would be half the above value, and the footing could be loaded to $(2)(5300) = 10,600$ psf. As the settlement of a rigid footing is between that at the center and corner of a flexible footing, a bearing pressure between 5300 and 10,600 psf would cause about 1 inch of settlement. Conservatively using 6000 psf, the load that could be applied to a 4 ft $\times$ 4 ft footing is

$$Q = 6000(4)(4) = 96,000 \text{ lb} = 96 \text{ kips}$$

The above answer could vary considerably depending on the selection of E_s and the rigidity of the footing.

15.17   (Soil from Example 15.13 with a 4 $\times$ 4 ft footing with a net load of 2 kips /ft^2). The solution uses the same equations as the example.

Top layer (midpoint 4.5 ft below the footing):

$$\Delta\sigma = (2000)(4)(4) \, / [(8.5)(8.5)] = 443 \text{ psf}$$

$$\sigma' + \Delta\sigma = 644 + 443 = 1087 \text{ psf}$$

Since 1087 is below the preconsolidation pressure of 1500 psf, the recompression index C_r is used.

$$\varepsilon_v = (1 \, / 1.9) \, [\, 0.04 \log (1087 \, / 644) \,] = 0.0048$$

$$S_{top} = 60 \text{ in } (0.0048) = 0.287 \text{ in}$$

Middle layer:

$$\Delta\sigma = (2000)(4)(4) \, / [(13.5)(13.5)] = 176 \text{ psf}$$

$$\sigma' + \Delta\sigma = 932 + 176 = 1108 \text{ psf}$$

$$\varepsilon_v = (1 \, / 1.9) \, [\, 0.04 \log (1108 \, / 932) \,] = 0.0016$$

$$S_{mid} = 60 \text{ in } (0.0016) = 0.096 \text{ in}$$

Bottom layer:

$$\Delta\sigma = (2000)(4)(4) \, / [(18.5)(18.5)] = 94 \text{ psf}$$

$$\sigma' + \Delta\sigma = 1220 + 94 = 1314 \text{ psf}$$

$$\varepsilon_v = (1 \, / 1.9) \, [\, 0.04 \log (\, 1314 \, / 1220 \,) \,] = 0.00068$$

$$S_{bot} = 60 \text{ in } (0.00068) = 0.041 \text{ in}$$

The total settlement is the sum of the layer settlements:

$$S = 0.287 + 0.096 + 0.041 = 0.42 \text{ inches}$$

15.18   Settlement versus time is evaluated using Equation 15.6.7. First, determine the time to fifty percent of the total settlement. From Table 15.3, The average degree of consolidation U_{avg} is 50% when the time factor T is 0.197. Since the 15 ft clay layer has sand above and below it, it is doubly drained and $H = 15/2 = 7.5$ ft. Rearranging Equation 15.6.7, the time for 50% consolidation is

$$t_{50} = T_{50} \, H^2 \, / c_v$$
$$= 0.197(7.5^2) \, / 0.2 = 55 \text{ days}$$

Similarly

$$t_{20} = T_{20} \, H^2 \, / c_v$$
$$= 0.031(7.5^2) \, / 0.2 = 9 \text{ days}$$
$$t_{90} = T_{90} \, H^2 \, / c_v$$
$$= 0.848(7.5^2) \, / 0.2 = 239 \text{ days}$$

Times for other degrees of consolidations can likewise be determined. The settlements at theses times are obtained by multiplying U_{avg} by the total predicted settlement:

At 9 days,

$$S = 0.20(0.42) = 0.08 \text{ in}$$

At 55 days,

$$S = 0.50(0.42) = 0.21 \text{ in}$$

At 239 days,

$$S = 0.90(0.38) = 0.34 \text{ in}$$

15.19   From Table 15.4, α is taken as 0.40 for $c = 3000$ psf. From Equation 15.7.5, the unit side resistance is

$$f_s = \alpha c = 0.4(3000) = 1200 \text{ psf}$$

The perimeter of the pile is $16\pi/12 = 4.19$ ft. The surface area of the pile is

$$(40 \text{ ft})(4.19 \text{ ft}) = 168 \text{ ft}^2$$

Multiplying the unit side resistance by the surface area, the total side capacity Q_{side} is

$$1200(168) = 201{,}600 \text{ lb}$$

The tip area of the pile is

$$A_{tip} = (1/4)\pi \ (16/12)^2 = 1.40 \text{ ft}^2$$

From Equation 15.11.1, the total tip capacity is

$$Q_{tip} = 9 \ (1.40)(3000) = 37{,}800 \text{ lb}$$

The total ultimate pile capacity is the sum of the side and tip capacity:

$$Q_{ult} = 201{,}600 \text{ lb} + 31{,}800 \text{ lb} \approx 240 \text{ kips}$$

The recommended allowable working load is obtained by applying a factor of safety of 3 or more. Using $FS = 3.0$,

$$Q_a = 80 \text{ kips}$$

15.20   From Equation For $\phi = 32°$, $K_a = \tan^2 (45 - 32/2) = 0.307$. The driving force is the active earth force. From Example 15.16, the active earth force for a sand backfill with no water table present is

$$P_a = 0.5 \ K_a \gamma H^2$$
$$= 0.5 \ (0.307)(125)(8^2) = 1228 \text{ lb /ft}$$

From Example 15.21, the resisting force is $W \tan \phi = 0.625 \ W$, where W is the weight of the wall. For a factor of safety of 2.0 against sliding, the resisting force must be twice the driving force, or $2(1228) = 2456$ lb /ft. Setting the resisting force to this value,

$$0.625 \ W = 2456$$
$$W = 3930 \text{ lb /ft}$$

Taking the density of concrete as 150 lb /ft^3, the weight of the wall can be expressed as

$$3930 = (0.5)(2+B)(8)(150)$$

Then

$$B = 4.55 \text{ ft}$$

For practicality, recommend a base width of 4.5 or 5.0 ft. For design, the wall should also be checked for overturning and should be founded below the frost depth.

15.21 First check the value of $\gamma H/c$:

$$125(12) / 3000 = 0.5$$

As this value is below 4.0, the clay is a stiff clay. From Equation 15.10.5, the design pressure over the middle half of the height is

$$p = 0.3 (125)(12) = 450 \text{ psf}$$

The design pressure diagram increases linearly from zero to the above value in the top quarter of the wall, and decreases linearly to zero in the bottom quarter. Hence the design pressure diagram is as shown below.

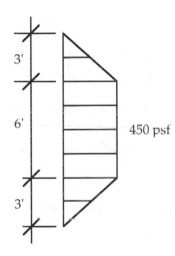

The total design earth force per foot of wall is

$$(0.5)(3)(450) + (6)(450) + (.5)(3)(450) = 4050 \text{ lb /ft}$$

Multiple Choice Problems (PE-Format)

15.22 e) From Equation 14.10.1, the total vertical stress at point A′ is

$$\sigma_v = \gamma H = (125 \text{ lb/ft}^3)(16 \text{ ft}) = 2000 \text{ lb/ft}^2$$

As there is no groundwater present, the pore pressure is zero, and the effective vertical stress is equal to the total vertical stress.

From Equation 15.8.3, the active earth pressure coefficient is

$$K_a = \tan^2 (45 - 34/2) = 0.2827$$

From equation 15.8.1, the horizontal effective stress σ'_h, which is the same as the horizontal active earth pressure p_a, is

$$\sigma'_h = K_a \sigma'_v = (0.2827)(2000 \text{ lb/ft}^2) = 565.4 \text{ lb/ft}^2$$

15.23 b) From Example 15.16, the total active earth force is

$$P_a = 0.5\ \gamma K_a H^2 = (0.5)(125 \text{ lb/ft}^3)(0.2827)(16 \text{ ft})^2 = 4523.2 \text{ lb /ft of wall.}$$

15.24 c) From Equation 15.8.6, the passive pressure coefficient for the material at the toe is

$$K_p = \tan^2 (45 + 30/2) = 3.000$$

Similar to problem 15.22, the effective vertical stress at the toe is

$$\sigma'_v = \gamma z - u = (120)(4) - 0 = 480 \text{ lb/ft}^2$$

The effective horizontal stress σ'_h , which is equal to the passive earth pressure is then

$$\sigma'_h = K_p\ \sigma'_v = (3.00)(480) = 1440 \text{ lb /ft}^2$$

15.25 d) Similar to problems 15.23, the total passive earth force is

$$P_p = 0.5\ \gamma K_p H^2 = (0.5)(120 \text{ lb/ft}^3)(3.000)(4 \text{ ft})^2 = 2880 \text{ lb /ft of wall.}$$

15.26 e) Following Example 15.21 and taking the unit weight of concrete to be 150 lb/ft^3, the weight of the free body above the base is the weight of the concrete plus the weight of the soil above the heel and the toe:

$$
\begin{aligned}
W &= W\text{soil heel} + W\text{soil toe} + W\text{base} + W\text{stem} \\
&= (6)(14)(125) + (3)(2)(120) + (11)(2)(150) + (2)(14)(150) \\
&= 10{,}500 + 720 + 3{,}300 + 4{,}200 \\
&= 18{,}720 \text{ lb /ft}
\end{aligned}
$$

The resisting normal force acting upward on the base is equal to this value.

The maximum base shear force is then

$$S = W \tan \phi = (18720 \text{ lb/ft})(\tan 30°) = 10,808 \text{ lb /ft}$$

15.27　d)　Following Example 15.21, the factor of safety against sliding is

$$FS = \frac{W \tan \phi + P_p}{P_a} = \frac{10808 + 2880}{4523} = 3.03$$

15.28　a)　Following Example 15.21 and using the forces from problem 15.26 and 15.25, the resisting moment is

$$
\begin{aligned}
M &= M\text{soil heel} + M\text{soil toe} + M\text{conc base} + M\text{conc stem} + M \text{ passive} \\
&= (10520)(8) + (720)(1.5) + (3300)(5.5) + (4200)(4) + (2880)(4/3) \\
&= 84,160 + 1080 + 18,150 + 16,800 + 3,840 \\
&= 124,030 \text{ ft-lb /ft of wall}
\end{aligned}
$$

The overturning moment is

$$
\begin{aligned}
M_O &= M_\text{active} \\
&= (4523.2 \text{ lb/ft})(16/3 \text{ ft}) \\
&= 24,123 \text{ ft-lb /ft of wall}
\end{aligned}
$$

The factor of safety is then

$$FS = M_r / M_o = 124,030/24,123 = 5.14$$

15.29　c)　The difference between the resisting moment and the overturning moment must be provided by the moment due to the resisting normal force on the base. The moment due to this force is then

$$M = 78,860 - 24,123 = 54,737 \text{ ft-lb/ft}$$

This moment is due to the normal force of 18,720 lb/ft acting at a moment arm of $x$bar from the toe. The moment arm is then

$$x\text{bar} = 54,737/18,720 = 2.92 \text{ ft}$$

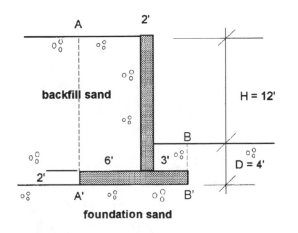

Engineering Economics

by Frank Hatfield

Chapter 16

Solutions to Practice Problems

16.1 **d)** Prestige.

16.2 **d)** $1000 \times 1.06^3 = \$1191$.

16.3 **c)** $150 / 1.08^2 = \$129$.

16.4 **d)** $F = 12,000(F/P)_3^{12} - 4000 = \$12,860$.

16.5 **c)** $P = 100,000(P/A)_\infty^8 = \$1,250,000$.

16.6 **b)** $500(P/A)_{12}^8 = 500 \times 7.536 = \$3768$.

16.7 **b), d)** $A:\ 300,000 + 35,000(P/A)_{30}^8 - 50,000(P/F)_{30}^8 = \$689,000$

$B:\ 689,000(A/P)_{30}^8 (P/A)_\infty^8 = \$765,000$.

16.8 **b)** $\left[1000(P/F)_5^{10} + 2000(P/F)_{10}^{10} + 3500(P/F)_{15}^{10}\right](A/P)_{20}^{10} = \$262$.

16.9 **b)** $10 \times 10^6 = X + 0.1 \times 10^6 (P/A)_\infty^6$. $\therefore X = 8.33 \times 10^6$.

16.10 **c)** $P = 6000(P/A)_3^{10} - 2000(P/G)_3^{10} = 10,300$.

16.11 **e)** By inspection.

16.12 **c)** $P = 200\left(P/A\right)_{10}^{6} = \$1472$.

16.13 **b)** $P = 12,000 + 500\left(P/A\right)_{3}^{10} + 1000\left(P/G\right)_{3}^{10} = \$15,570$.

16.14 **e)** $2P = \left(F/P\right)_{5}^{i} P$. $\left(F/P\right)_{5}^{i} = (1+i)^{5} = 2$. $\therefore i = 15\%$.

16.15 **d)** $54,000\left(A/P\right)_{30}^{8} = \$4800$.

16.16 **c)** $10,000\left(F/P\right)_{4}^{6} - 3000 = \$9625$.

16.17 **a)** $18,000\left(A/F\right)_{8}^{4} = \$1953$.

16.18 **a)** $1000\left(P/A\right)_{\infty}^{10} + 1000\left(P/A\right)_{10}^{10} = \$16,145$. Note: for the first 10 years, this accounts for $2000/yr.

16.19 **a)** $20,000\left(P/A\right)_{6}^{8}\left(P/F\right)_{1}^{8} = \$85,600$.

16.20 **b)** $e^{0.1} - 1 = 0.10517$ or 10.517%.

16.21 **b)** $675 = 50 + 30\left(P/A\right)_{24}^{i}$. $\therefore \left(P/A\right)_{24}^{i} = 20.833$.
$\therefore$ by trial and error $i = 0.0116$. $\therefore 12i = 0.139$ or 13.9%.

16.22 **b)** $500,000 + \left[200,000 + 100,000(A/P)_{3}^{4}\right](P/A)_{\infty}^{4} = \$6,300,000$.

16.23 **d)** $400 = 35 + 45\left(P/A\right)_{10}^{i}$. $\left(P/A\right)_{10}^{i} = 8.11$. $i = 4\%$. $i_{n} = 12i$. $\therefore i_{n} = 48\%$.

16.24 **e)** From preceding solution $i = 4\%$.
$i_{e} = (1+0.04)^{12} - 1 = 0.601$. $\therefore i_{e} = 60.1\%$.

16.25 **d)** $1000 = 300\left(P/A\right)_{4}^{i}$. $\left(P/A\right)_{4}^{i} = 3.33$. $i = 7.7\%$. $\therefore i_{n} = 52 \times 7.7 = 400\%$.

16.26 **a)** $A: -16,000\left(A/P\right)_{8}^{12} - 2000 + 2000(A/F)_{8}^{12} = -\$5058$.
$B: -30,000\left(A/P\right)_{15}^{12} - 1000 + 5000(A/F)_{15}^{12} = -\$5270$

16.27 **d)** $P\left(A/P\right)_{20}^{10} = \left[30,000 + 10,000\left(P/F\right)_{4}^{10}\right]\left(A/P\right)_{8}^{10}$. $\therefore P = \$58,760$.

16.28 **b)** $120,000 + 9000\left(P/A\right)_{6}^{10} - 25,000\left(P/F\right)_{6}^{10} = \$145,000$.

16.29 **e)** $\left[100,000(A/P)_{10}^{10} + 10,000\right]\left(P/A\right)_{\infty}^{10} = \$262,700$.

16.30 **d)** $A: 50,000 + 800\left(P/A\right)_{20}^{8} = \$57,900$
$B: 30,000 + 500\left(A/P\right)_{20}^{8} + \left[30,000 + 400\left(P/A\right)_{10}^{8}\right]\left(P/F\right)_{10}^{8} = \$50,000$.

16.31 **b)** $A: \ -116 + 0.93 \times 206 \left(P/A \right)_8^{10} = \$906$

$B: \ -60 + 0.89 \times 206 \left(P/A \right)_8^{10} = \$918$

16.32 **b)** $A: \ \left[-25{,}000 + 83{,}000 \left(P/F \right)_3^{10} \right] \left(A/P \right)_6^{10} - 6000 + 13{,}000 \left(A/F \right)_6^{10} = \$4260.$

16.33 **d)** $(20{,}000 + P)\left(A/P \right)_9^8 - 300 - 5000 \left(A/F \right)_9^8 = 20{,}000 \left(A/P \right)_6^8 - 5000 \left(A/F \right)_6^8.$

$\therefore P = \$7140.$

16.34 **b)** $(23{,}000 - 15{,}000) + (23{,}000 - 32{,}500)\left(P/F \right)_N^6 = 0. \quad (1.06)^{-N} = 0.84.$

$\therefore N = 3 \text{ yrs}.$

16.35 **b)** $-3500 \left(A/P \right)_5^8 + 12(200 - 50) = \$923.$

16.36 **c)** $5000 + 1.50n = 4.00n. \quad \therefore n = 2000.$

16.37 **b)** $40{,}000x = 500{,}000\left(A/P \right)_{15}^8 - 100{,}000 \left(A/F \right)_{15}^8 + 30{,}000x.$

$\therefore \ x = 5.47. \quad \text{Use } x = 6.$

16.38 **d)** $1200 + 40(20 + x) = 90(20) + 50x. \quad \therefore \ x = 20$

16.39 **b)** $65(50) - 65(40) - 1200 = -550$

16.40 **d)** $4500 \left(A/P \right)_{10}^8 + (18.7 + .1x)75/.9 = 3000 \left(A/P \right)_{10}^8 + (18.7 + .1x)75/.89 .$

$\therefore \ x = 2200 \text{ hr.}, \ B$

16.41 **d)** $10{,}000 - (10{,}000 - 1300)8/12 = 4200$

16.42 **c)** $D = (1500 - 1500)/3 = 1000$

16.43 **a)** $(300{,}000 - 50{,}000)/7 = 35{,}714$

$300{,}000 - 4(35{,}714) = 157{,}143$

16.44 **c)** $(300{,}000/7)0.5 = 21{,}419$

$300{,}000/7 = 42{,}857$

$300{,}000 - 3.5(42{,}857) = 150{,}000$

16.45 **e)** $300{,}000 \times 2/7 = 85{,}714$

$(300{,}000 - 85{,}714)2/7 = 61{,}224$

$300{,}000 - 85{,}714 - 61{,}224 - 43{,}732 - 31{,}237 = 78{,}092$

16.46 **d)** $(300{,}000 \times 2/7)0.5 = 42{,}857$

$(300{,}000 - 42{,}857)2/7 = 73{,}469$

$300{,}000 - 42{,}857 - 73{,}469 - 52{,}478 - 37{,}484 = 93{,}711$

16.47 **a)** $P = 400S - (35000 + 0.5S^2). \quad S = 1000 \text{ gives } P = -\$135{,}000$

16.48 **b)** $P = 0.\ S = \left(-400 \pm \sqrt{400^2 - 4 \times 0.5 \times 35000} \right) \Big/ (-2 \times 0.5).$

   $\therefore S = 400 \pm 300$ units/mo

16.49 **e)** $dP/dS = 400 - 2 \times 0.5S = 0.\ \ S = 400.\ \ P = \$45,000$ per month.

16.50 **c)** $U = P/S = 400 - (35000S^{-1} + 0.5S).\ \ dU/dS = 35,000S^{-2} - 0.5 = 0.$
   $S = 265.\ \ U = \$135/$unit.

16.51 **c)** $\text{ANEV} = \left[-4000(A/P)^{20}_5 - 500 \right] - [-1500] = -\$338.$

16.52 **d)** $\text{ANEV} = 0 = \left[-4000(A/P)^{20}_5 - 5X \right] - [-15X].\ \ \therefore X = 134$ kg/yr.

16.53 **b)** $D = 4000/3 = \$1333.\ V_2 = 4000 - 2 \times 1333 = \$1333.$

16.54 **b)** $P = 20 \times (0.50 - 0.05) + 15(0.25 - 0.05) - 25 = -\$13.$

16.55 **c)** $P = (20 + 15) \times (0.25 - 0.05) - 25 = -\$18.$

16.56 **d)** $D = \dfrac{(90,000 - 15,000)}{5} = 15,000;\ \ V_2 = 90,000 - 2 \times 15,000 = \$60,000$

16.57 **d)** $D_1 = \dfrac{90,000 \times 2}{5} = 36,000.\ \ V_1 = 90,000 - 36,000 = 54,000$

   $D_2 = \dfrac{54,000 \times 2}{5} = 21,600.\ \ V_2 = 54,000 - 21,600 = 32,400$

   $D_3 = \dfrac{32,400 \times 2}{5} = 12,960.\ \ V_3 = 32,400 - 12,960 = 19,440$

   $D_4 = \dfrac{19,440 \times 2}{5} = \ \ 7,776.\ \ V_4 = 19,440 - \ \ 7,776 = 11,664 < 15,000$

16.58 **c)** $\text{ANEV} = -(170,000 - 90,000)(A/P)^{20}_5 - (70,000 - 44,000)$

   $+ (50,000 - 15,000)(A/F)^{20}_5 = \ -\$48,000$

16.59 **d)** $\text{PNEV} = (160,000 - 100,000)(P/A)^{20}_5 - (170,000 - 90,000)$

   $-(70,000 - 44,000)(P/A)^{20}_5 + (50,000 - 15,000)(P/F)^{20}_5 = \$36,000.$

16.60 **a)** $dC/dt = I - \left(L/t^2 \right) = 0;\ \ t = \sqrt{L/I}.$

Use the 5-Step PE Prep System to Pass!

 ✔ **$20 Off System Package!**

All-in-one volume!
Only $69.95!

Step #1: Get the Best PE Review!

○ ***Principles & Practice of Electrical/Civil/Mechanical Engineering***
Edited by Merle C. Potter, PhD, PE (GLP), these highly-effective PE reviews are from the publisher of the best FE title. To create your own 'Ultra PE Prep System', simply order a review, a sample exam and a discount Handbook and you have all the resources you need. So, start with a great ***review***... These are the ___only___ titles that concisely cover all critical aspects of the PE. Approx. 600 pp. (Solutions manuals to problems available separately.)

Only $69.95!

Step #2: Select a Sample Exam!

Only $69.95!

○ ***Official NCEES Sample PE Exams—$32 each.*** All subjects available. (Structural I and II, $48.) After your initial prep, just sit and take your sample exam (open-book) to verify readiness or identify weaknesses. Use your review and handbook to assist you—you have easy access to all necessary information.

Subjects desired: _____

Step #3: Order a Handbook!

Reg. $120 **Only $99!**

○ ***Mark's Standard Handbook for Mechanical Engineers.***
By E. Avallone (10th ed., McGraw-Hill). The book to turn to for practical advice and quick answers on all ME principles, standards and practices. 1,792 pages.

○ ***Standard Handbook for Civil Engineers.***
By F. Merritt (4th ed., McGraw-Hill). New edition of the most thorough compilation of facts and figures in all areas of civil engineering. Best guide, best buy, and best-seller since 1968! 1,456 pp.

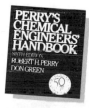

Only $99! Reg. $125

○ ***Perry's Chemical Engineer's Handbook.***
By R. Perry (7th ed., McGraw-Hill). The best info on all aspects of ChemE, all in one place! The standard for over 40 years! 2,640 pages.

Reg. $115 **Only $99!**

○ ***The Electrical Engineering Handbook.***
Edited by Richard C. Dorf, PhD (2nd ed., CRC Press). The premier EE reference. All-new handbook with up-to-date info in all areas of EE. A must for practicing engineers! 2,696 pages.

Only $99! Reg. $125

Step #4: Don't Forget a Test-Beating Calculator!

Hewlett-Packard
G Only $120!
GX Only $230!
G Reg. $165
GX Reg. $285

○ ***HP 48G(X) Programmable Calculator.*** Most states allow sophisticated calculators to be used when taking PE & FE exams. HP-48 has hundreds of built-in equations you will need. GX plug-in card offers exact FE/PE equations! ***JumpStart the HP 48***, written for engineers, is a handy GLP title (new 2nd ed.) that's superior to HP's manual.

Name (print): _____

Street Address: _____

CP-98

Step #5: Fill out this Quick'n'Easy ORDER FORM

Credit Card type: VISA   MC   (circle one)

Number: _____ Exp.: _____

30-day money-back guarantee on GLP titles only.
Mail this form and check or money order to:
Great Lakes Press POB 550 Wildwood, MO 63040-0550

Call 1-800-837-0201 / Fax 1-314-273-6086
online ordering: www.glpbooks.com
email: order@glpbooks.com

Name of Title	Price	Quan.	Multi-Discount!	Subtotal
PE Review EE/CE/ME	$69.95	X —	**$10 off if ordering PE & Sol'ns!**	=
Solutions EE/CE/ME	$19.95	X —		=
CE Handbook	$99	X —		=
ME Handbook	$99	X —		=
EE Handbook	$99	X —	**Order from Steps 1, 2 & 3 $20 off!**	=
ChemE Handbook	$99	X —		=
NCEES PE Samples	$32	X —		=
Calculator G/GX	$120/230	X —	N.A.	=
GX Card for FE/PE	$109.95	X —	N.A.	=
Jump Start HP-48	$19.95	X ↓	N.A.	=
Shipping per Item	$5	X	➤	=

(MI & MO residents add 6% sales tax.) TOTAL $ _____

Review Course Videos for FE & PE Exams

Now available!
Call for more info and pricing!

✔ **For Sale or Rent!**

PE Civil Engineering Video Course

◯ *From ASCE. 12 tapes, 23 hours total.*

FE/EIT Civil Engineering Video Course

◯ *From ASCE. Available soon!*

Interested? Check the circles! SEND FOR FREE INFO!

Name (print): _____

Address: _____

_____

_____

Mail or fax this form to ·
Great Lakes Press POB 550
Wildwood, MO 63040-0550

Call 1-800 837-0201 / fax 1-314-273-6086
www.glpbooks.com / custserv@glpbooks.com

CP-98

$10 Gift Certificate

Snip and save this coupon for $10 off your next purchase of GLP titles (on orders over $50).

This coupon is transferable, so feel free to share it with a friend!

CP-98